The solid–liquid interface

Cambridge Solid State Science Series

EDITORS:
Professor R. W. Cahn
Applied Sciences Laboratory, University of Sussex
Professor A. J. Forty
Department of Physics, University of Warwick
Professor I. M. Ward
Department of Physics, University of Leeds

D. P. WOODRUFF

Lecturer in Physics, University of Warwick

The solid–liquid interface

CAMBRIDGE UNIVERSITY PRESS

CAMBRIDGE UNIVERSITY PRESS
Cambridge, New York, Melbourne, Madrid, Cape Town, Singapore,
São Paulo, Delhi, Dubai, Tokyo, Mexico City

Cambridge University Press
The Edinburgh Building, Cambridge CB2 8RU, UK

Published in the United States of America by Cambridge University Press, New York

www.cambridge.org
Information on this title: www.cambridge.org/9780521299718

© Cambridge University Press 1973

First published 1973
Re-issued 2010

A catalogue record for this publication is available from the British Library

Library of Congress Catalogue Card Number: 72–91362

ISBN 978-0-521-20123-0 Hardback
ISBN 978-0-521-29971-8 Paperback

Contents

Preface

This book sets out to discuss the state of our knowledge of the solid–liquid (i.e. the solid–melt) interface. While the first objective is to understand the structure or properties of this interface in an equilibrium state, most experiments have been concerned with states in which the interface is moving, and most particularly solidification. The relatively recent need to be able to grow well-defined single crystals of many materials (particularly for semiconductor devices), often by growth from the melt, has generated research which has provided a much better understanding of the fundamental processes of this mode of crystal growth. At the same time, it has been appreciated that an improved understanding of these processes can help in understanding and controlling the structures of quite massive castings of metals.

Books already exist on the metallurgical applications of controlled solidification, and on the many specialised and complex forms of the melt growth technique applied to single crystal growth. Instead, this book attempts to cover the basic background physics associated with solidification and melting. As such it will not tell the reader how to go out and grow a single crystal or make a better casting; but it may allow him to understand better why certain precautions and methods are better than others in achieving this aim. The title of *The Solid–Liquid Interface* was chosen as indicating the viewpoint on the subject to be adopted. For example, the basic thermodynamics of solid surfaces is discussed and our rather limited knowledge of the interface in equilibrium is reviewed; regrettably, little is known in detail about the interfacial free energy of the solid–liquid interface and, more particularly, its anisotropy, and so its role in determining crystal growth behaviour has yet to be properly assessed. Another topic which is dealt with here in much more detail than in conventional texts on solidification is that of interfacial instability. The extremely complex shapes of dendritic growth forms encountered in castings originate because these forms are the simplest ones that are stable, and our understanding of the detailed fashion in which simpler shapes become unstable has greatly improved in the last few years. Finally, wherever possible, not only solidification, but also melting, has been discussed. The practical applications of our understanding of solidification are far more extensive than of melting, but the two processes are closely related and an understanding of one can supplement an understanding of the other.

I should like to thank all those authors who kindly provided me with photographs from their publications, and elsewhere, which I have used in this book, and most particularly to Dr J. D. Hunt of the University of Oxford who not only provided me with many such illustrations but also offered helpful comments after an initial reading of the chapter on eutectic growth.

D.P.W.

July 1972

1   Interfacial free energy and the γ-plot

1.1   The free energy of an interface

Whilst the interest in the atomic structure and electronic behaviour of surfaces has arisen relatively recently, the method of treating interfaces thermodynamically was introduced as long ago as the last century. Gibbs, in particular, published a considerable contribution to the field in 1878 and much of this work still forms the basis of a large part of the present understanding of interfacial phenomena.

Of particular interest is the concept of the free energy, per unit area, of an interface, and necessarily associated with this, the method of defining the interface position relative to the two phases which lie on either side of it. The concept of a free energy associated with an interface may easily be appreciated by considering the total free energy of some system consisting of two phases of volumes V_1 and V_2 in contact. The total Gibbs free energy, for instance, will be given by

$$G = G_1 V_1 + G_2 V_2 + \gamma_{12} A_{12} \tag{1.1}$$

where A_{12} is the area of the interface separating the phase and γ_{12} is the Gibbs free energy per unit area of the interface. An understanding of a surface energy may be achieved by considering the surface of a material whose energy may be described on a pairwise bonding model. Clearly, because of the different environment of atoms near the surface of the material, the energy of these atoms will be different from those in the bulk. For the sake of convenience this difference is associated with the dividing surface between the phases.

The Gibbs free energy per unit area of a surface may be decomposed, in the normal way, as

$$\gamma = \epsilon - T\eta + pv \tag{1.2}$$

where ϵ is the surface energy per unit area, η the surface entropy per unit area and v the surface volume per unit area. This final term, v, is associated with the change in atom density in a material near the interface; for instance, there may be a different spacing of the outermost planes near a free solid surface. This is an extremely small term and may be neglected in almost all cases of interest. Thus, in general, the Gibbs free energy per unit area is equivalent to the Helmholtz free energy per unit area which is also often defined as γ. A more complete treatment and derivation of the concept of surface free energy was given by Gibbs (1878) but the above discussion will suffice for our present needs. It is, however, of interest to point out that Gibbs accounted for adsorption at the interface by the addition of terms $\mu_i \Gamma_i$ to the surface free energy where μ_i is the chemical potential of the ith species of adsorbed atoms, and Γ_i is the surface density of these atoms. Also, in general, the *Gibbs dividing plane* or interface, is defined only as a plane passing through all points having a similar environment in the boundary region between the two phases. This surface has a degree of freedom in displacement perpendicular to the surface; this is removed for one-component systems by the necessity that the surface mass term shall be zero. Thus, if an equation of the general form of (1.1) is written with the total mass of the system on the left hand side and the G_1, and G_2 replaced by the densities of the two bulk phases, the surface is positioned such that no surface term need appear in this equation. For multi-component systems, the problem of defining the position and thickness of the interface has been discussed in detail by Cahn & Hilliard (1958).

1.2   Surface tension and its relation to surface free energy

The definitions given above for the interfacial free energy per unit area (which will be referred to, for the sake of convenience, as the surface free energy) corresponds to the work involved in creating unit area of interface. In the rest of the discussion in this chapter it is this quantity which is of significance in determining the equilibrium in a two phase system (when the total Gibbs free energy is a minimum). However, as will be seen in the next chapter, the quantity measured experimentally is usually the surface tension of the interface. For an interface between two fluids the relationship between these two quantities is simple. In this case a consideration of the work done against the surface tension in expanding an interface to create new area of interface shows that this work is equal to the surface free energy of the same area of new interface. Thus the surface tension and surface free energy are numerically equal if expressed in similar units. However, this is not necessarily true for an interface which is bounded on at least one side by a solid (which for convenience will be referred to as a solid surface, though this may be a free solid surface, a solid–liquid interface or a solid–solid interface). As Gibbs pointed out, whilst the surface free energy corresponds to the work done in *forming* the surface, the surface tension depends on the work done in *stretching* the surface. The equivalence of these two processes requires the tacit assumption that there is no marked change in surface structure associated with the latter process and thus no work is spent in deforming the surface. In using experimental determinations of the surface tension to give a value of the surface free energy it is assumed, therefore, that the temperature at which the experiments are performed is sufficiently high for surface and bulk diffusion to correct the distortion effects and so permit the equality to be used.

1.3   The γ-plot and the terrace-ledge-kink model of a surface

In order to fully characterise an interface in terms of its surface free energy, γ must be known for all orientations of the interface and conventionally it is plotted on a polar diagram such that the length of the radius coordinate is proportional to the value of γ for a surface perpendicular to the direction of the radius vector. For a simple liquid interface there is no orientation dependence of γ and so the γ-plot (as this polar diagram is commonly known) is a sphere, and any section passing through the origin of the coordinate system (which therefore provides a more convenient two-dimensional γ-plot) is a circle.

For a solid surface, however, γ is clearly a function of crystallographic orientation and so the γ-plot will not be spherical. To determine the general features of the γ-plot for a solid surface it will be convenient to consider the free surface of a solid at the absolute zero of temperature so as to remove thermal disordering and entropy effects. Let us first consider surfaces having orientations close to that of a low index plane. These surfaces will appear as a series of terraces of the low index plane, the step density in this terraced structure being characteristic of the deviation of the surface normal from that of the low-index plane, θ. Thus for such a surface composed of steps, or ledges of height a, having a mean separation in the complex surface, l (see fig. 1.1(a)), it is clear that

$$\sin \theta = a/l. \tag{1.3}$$

Now according to the terrace-ledge-kink (TLK) model of a surface, the surface energy can be decomposed into terms giving the energy for the low index plane, γ_0 per unit area of the low index plane (the terraces), the ledge energy, β per unit length of ledge, and in the case of more complex orientations than that considered here there will be a further term due to the energy of kinks in the ledges (fig. 1.1(b) shows an example of such a surface). For any particular azimuth the kink density on the ledges will be constant, and only the separation of the ledges will change with θ, so the particular model used here of a low

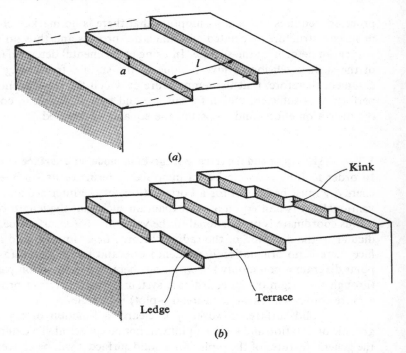

Fig. 1.1. (*a*) A simple stepped or ledged surface with low index terraces. (*b*) More complex surface including kinked ledges.

index azimuth involves no loss of generality providing that the absolute value of the ledge energy is not specified. For this surface, therefore, as shown in fig. 1.1, the surface energy per unit area is given by

$$\gamma_\theta = \gamma_0 \cos |\theta| + \beta/l$$

$$= \gamma_0 \cos |\theta| + \frac{\beta}{a} \sin |\theta|. \tag{1.4}$$

Note that the $|\theta|$ terms have been introduced because γ_θ is independent of the sign of θ due to the symmetry of the situation. It is easy to show that (1.4) describes the relationship shown in fig. 1.2. The point of particular interest is that the graph shows that there will be a 'cusp' in the γ-plot at orientations corresponding to low index planes. This cusp is a sharp minimum in γ_θ but is a mathematical singularity rather than a simple minimum (i.e. $\partial\gamma_\theta/\partial\theta$ is discontinuous at this point). It is not, in fact, a point of self-tangency which is the more correct mathematical definition of a cusp in the normal way, but the term is in popular use and usefully describes the appearance of the curve.

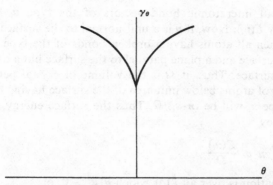

Fig. 1.2. The dependence of γ_θ on θ for the simple model depicted in fig. 1.1.

While this application of the TLK model of the surface is a very simple and therefore special case, it is possible, by suitable mental agility in describing a more complex surface on a TLK model, to show that cusps will appear in the γ-plot at all orientations corresponding to surfaces having rational Miller indices. Each of these singularities corresponds, as does a rational set of Miller indices, to a particular long range ordering described in the surface by particular sets of ledge and kink spacings. Evidently, in this model no account is being taken of thermal vibrations of the atoms which might destroy this long range order; this point will be discussed later in this chapter (§1.5).

It should, however, be pointed out that this treatment and the one which follows are designed to determine the *surface energy* which is only equivalent to the *free surface energy*, γ, at the absolute zero of temperature. However, this is the only temperature at which the above considerations are valid. Firstly, however, we shall use the other commonly applied method of determining surface energies, in its most general form to show more clearly the shape of the γ-plot for a solid surface.

1.4   Pairwise bonding models

This alternative model of a solid surface is derived from assuming that a crystal consists of atoms interacting attractively in pairs by means of forces of finite range. The surface energy of a surface may then be given by the sum of the energies of all bonds broken by the surface and passing through unit area of the surface. It is easy to see that this is the same as the sum of the energies of all bonds broken by the surface originating from the atoms situated below unit area of the surface.

First, consider all interatomic bond vectors of the type u_i having associated energy $E(u_i)$. Now, if n is a unit normal to the surface under consideration, then all atoms having broken bonds of the type u_i will lie between the surface and a plane parallel to the surface but a distance $n \cdot u_i$ below the surface. Thus, if Ω is the volume of crystal per atom then the number of atoms below unit area of the surface having broken bonds of the type u_i will be $(n \cdot u_i)/\Omega$. Thus the surface energy of this surface is given by

$$\gamma_n = \sum n \cdot u_i \frac{E(u_i)}{\Omega} \tag{1.5}$$

where the summation is over all i for which $n \cdot u_i > 0$, or

$$\gamma_n = n \cdot \sum_i u_i \frac{E(u_i)}{\Omega}. \tag{1.6}$$

Notice that bonds for which $n \cdot u_i < 0$ are directed back into the crystal and are therefore not broken.

Now if the interatomic forces are of finite range, this summation has a finite number of terms and also there is a small range of orientations for which any set of i (for which $n \cdot u_i > 0$) is constant; these orientations lie in a pyramid, p, say. Within this pyramid we may write

$$s_p = \sum_i u_i \frac{E(u_i)}{\Omega} \tag{1.7}$$

where s_p is a fixed vector in p, and so within p

$$\gamma_n = n \cdot s_p. \tag{1.8}$$

Thus, we have from this relation, that the projection of s_p on n has magnitude γ_n. That is, within the pyramid p, the locus of the end of the vector $n\gamma_n$ is a sphere passing through the origin of the polar (γ-plot) coordinate system. This is shown in fig. 1.3. In general, therefore, providing that a pairwise bonding model is a suitable description of the crystal and its surface, the γ-plot will be composed entirely of portions of spheres which pass through the origin. Moreover, the inter-sections of these spheres (where cusps will be formed) occur where the set of u_i for which $n \cdot u_i > 0$ changes; this occurs when one or more of the $n \cdot u_i = 0$, a condition which clearly corresponds to some rational index orientation. Thus the general picture of a solid surface γ-plot at absolute zero of temperature is of a surface composed entirely of portions of spheres (all of which will be convex outside as they pass through the origin), with cusps at the junctions of these spheres corresponding to all rational index orientations. This shape has been aptly described

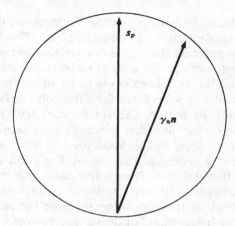

Fig. 1.3. Vector diagram for s_p and $n\gamma_n$ showing the section of the γ-plot generated in the pyramid p.

Fig. 1.4. A schematic γ-plot at 0°K.

by Frank as a 'raspberry' figure. A schematic γ-plot of this type is shown in fig. 1.4.

1.5   Effect of temperature on the γ-plot

At elevated temperature the shape of the γ-plot simplifies from that shape described above. This is because the increase in temperature results in a certain amount of surface disordering (for instance, in the form of the equilibrium concentration of kinks on ledges). This tends to destroy the long range order on the high index surfaces and so the singularities in the γ-plot corresponding to these surfaces may be expected to disappear and the associated cusps to be smoothed out. This problem has been treated in some detail for a free solid surface in the TLK model, particularly by Burton, Cabrera & Frank (1951). For a high index surface, the surface disorders gradually with increasing temperature as the density of opposite sign kinks on the surface ledges increases, until all long range order has disappeared. For a low index surface the picture is rather different. Disordering can occur by the generation of surface vacancies and adatoms but it is important to consider cooperative effects as the population of these increase. In analysing this situation the 'roughening' of the surface was considered on a two, three and five level model using Bethe's (1935) method for treating order–disorder phenomena. In this way they showed that the disordering of the surface occurs largely within quite a narrow temperature range after the fashion of a phase transition. For this reason the effect is known as 'surface melting'. Below the surface melting temperature the surface is well-ordered and smooth on an atomic scale. Frank (1958) has suggested that an interface of this type be called a 'singular' surface. Above the transition temperature, the surface atoms are in a completely disordered state and thus the singularity in the γ-plot corresponding to that orientation of surface will disappear, and the cusp will be rounded initially into a simple minimum. Burton, Cabrera and Frank also made some semi-quantitative predictions as to the value of the surface melting temperature for different surfaces, and found that for low index surfaces corresponding to planes of close packing in which the atoms are bound within the plane by bonds in two directions, the surface melting temperature is likely to be significantly higher than the bulk melting point. Thus surface melting will not be observed on these planes. For other orientations, however, surface melting is likely to occur below the bulk melting temperature. These results show that the complex γ-plot shape predicted in §1.4 for absolute zero temperature will be simplified by a rounding-off of many of the cusps at elevated temperatures. The extent to which this process has occurred depends on the temperature but in general there will always be some cusps remaining right up to the melting point of the solid. Fig. 1.5 depicts schematically the γ-plot of a solid surface at some elevated temperature.

Since the work of Burton, Cabrera and Frank other authors have shown that different models of the surface predict similar effects.

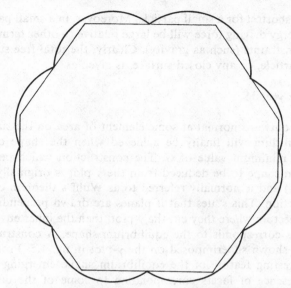

Fig. 1.5. A schematic γ-plot at elevated temperature. The Wulff construction for the equilibrium shape of a particle is also shown.

Mullins (1959) has shown that a very simple treatment using Bragg–Williams order–disorder theory predicts the surface melting type of behaviour. Gruber & Mullins (1967) have considered the disappearance of γ-plot cusps for low index plane surfaces, taking into account the configurational entropy of the surface, and similarly predict the occurrence of melting and confirm its lack of importance for the close-packed plane orientations below the melting point of the bulk solid.

1.6   Equilibrium shape of a surface

From a complete knowledge of the γ-plot for the interface between two phases it is possible to determine the equilibrium shape of the interface. The most general case is that of the equilibrium shape of a particle of one phase surrounded by the other phase. In the absence of external constraints the equilibrium shape will not be a function of which phase is the 'particle' and which is the surrounding phase. For instance, the equilibrium shape of a solid particle surrounded by its own vapour will be the same as that of a void in the solid. Experimentally, one might start with a very small spherical particle and observe the change in shape as the surface approaches equilibrium. In principle, any particles of any size will eventually achieve thermodynamic equilibrium and the same equilibrium shape, but in practice the time taken to achieve equilibrium is shortest for a small particle. Moreover, in a small particle the surface energy driving force will be large relative to other terms due to external constraints (such as gravity). Clearly, the total free surface energy of a particle, or any closed surface, is given as

$$G = \int \gamma(\mathbf{n}) \, dS, \tag{1.9}$$

where $\mathbf{n}$ is the outward normal at some element of area on the surface dS, and equilibrium will finally be achieved when the shape corresponds to the minimum value of G. The construction which permits the equilibrium shape to be deduced from the γ-plot is originally due to Wulff (1901) and is normally referred to as Wulff's theorem or the Wulff construction. This states that if planes are drawn perpendicular to the radius vectors where they cut the γ-plot, then the inner envelope of these planes corresponds to the equilibrium shape. A construction of this type is shown superimposed on the γ-plot in fig. 1.5. Probably the most interesting feature of the equilibrium shape emerging from this is the presence of facets corresponding to some of the deepest cusps and therefore being of rather low index (close-packed plane) orientations. Thus a large number of orientations are not represented on the equilibrium surface as the portion of the γ-plot corresponding to these orientations falls outside the constructed equilibrium surface shape. Evidently, a crystal in equilibrium at absolute zero (could such a state be achieved) would have an entirely polygonal shape, being composed entirely of facets. At elevated temperatures this need not be true, however, because if the cusps which do produce facets are shallow, rounded regions may appear on the equilibrium shape. However, as long as some cusps remain in the γ-plot at least some facets will appear in the equilibrium shape. For the case of free solid surfaces, therefore, which were treated in the Burton, Cabrera and Frank analysis discussed in the previous section, some facets will be expected to remain on the equilibrium shape right up to the bulk melting point.

While equilibration over the overall shape of a particle, for the free solid surface at least, is an extremely slow process, the local equilibrium in a surface may be expected to occur in a much shorter time because of the smaller amount of mass transfer and shorter distances of transfer required to achieve the desired state. Further deductions can be made about the shape of such a surface from the γ-plot, for, if the surface has an orientation corresponding to a large surface free energy, it is evident that a breakdown into a hill-and-valley structure of lower free surface energy surfaces may be favourable, despite the higher surface area incurred by this change. It is this effect which leads to the so-called thermal faceting of solid surfaces at elevated temperatures; this is when a complex surface breaks up into a hill-and-valley structure of low index facets or low index facets plus short regions of complex surface. In these experiments the crystal is held at an elevated temperature for which the mass transport at the surface is sufficiently rapid to allow this change to take place in a reasonable period of time. In fact, it can be shown that if a surface corresponds to one which appears on the equilibrium shape (as determined by the Wulff construction) then it will be stable against changes of this kind, but in any other case there will exist a hill-and-valley structure of low total free energy which the surface will therefore try to adopt. The proof of these, and several other theorems relating to the interrelation of the equilibrium surface morphology and the γ-plot have been given by Herring (1951*a*). However, the above results are the main ones of interest and will suffice for the discussion of solid–liquid interfaces which follows.

General reading
An excellent brief review of the basic thermodynamics of surfaces is given in a paper by F. C. Frank ('The geometrical thermodynamics of surfaces', in *Metal Surfaces*, ASM Metals Park, Ohio (1963), p. 1), and the paper by Gibbs (1878) is of considerable historical interest in this context, though its content is at times rather involved for the immediate interest of the solid–liquid interface. The paper by Burton, Cabrera & Frank (1951) is of considerable importance in the field of solid surfaces and crystal growth, and Herring's (1951) paper contains simple and elegant proofs of several important theorems relating to γ-plots and equilibrium surfaces.

2 The experimental determination of the solid–liquid interfacial free energy γ_{SL}

2.1 Methods of interface intersections

One of the most general methods of measuring the surface tension of a surface or an interface involves the study of the configuration adopted by it where it intersects some other interface (or interfaces). By considering the angles of intersection of the interfaces when the system is in equilibrium, the relative values of the surface tensions acting may be evaluated. Fig. 2.1 shows the simplest example of such intersection of three interfaces between phases 1, 2 and 3. In a specific situation two of these phases may be the same; for instance one of the interfaces may be a grain boundary in a solid in which case the other two interfaces will be of the same type (solid–vapour interfaces, say, if we are considering the intersection of a grain boundary with a free surface). However, for simplicity all the phases will initially be considered to be simple fluids so that the γs are isotropic. In this case too, it is correct to equate the surface tension with the surface free energy, γ. This equality is assumed in all methods of interface intersections; the basis for such an assumption is explained in §1.2. It should be noted, however, that this is not always true for solid surfaces.

For the simple case depicted in fig. 2.1,

$$\frac{\gamma_{12}}{\sin c} = \frac{\gamma_{23}}{\sin a} = \frac{\gamma_{31}}{\sin b}. \qquad (2.1)$$

Thus, a knowledge of the angles of intersection of the interfaces, which may be measured directly, gives the relative values of the surface tensions associated with the interfaces. Prior knowledge of the absolute value of one of the surface tensions therefore results in a determination of the values of the others.

Although not usually applicable to solid–melt interface study, for reasons which will become clear later, one example of the application of this method is the 'sessile drop' technique which uses the equilibrium of a liquid drop on a solid surface (fig. 2.2). Note that in this case equilibrium will only be achieved when the solid and liquid are also in equilibrium with the surrounding vapour, though at normal temperatures the solid–vapour equilibrium is not likely to be an important factor. Strictly, the equilibrium situation is depicted by fig. 2.2(a), but

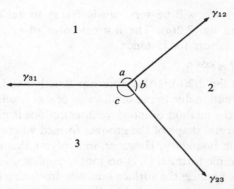

Fig. 2.1. Surface tensions acting at the intersection of three interfaces between phases 1, 2 and 3.

in practice this will only be achieved in a reasonable time for three fluid phases. Assuming that it is achieved, however, this state may be represented by the equations

$$\gamma_{31} = \gamma_{12} \cos \alpha + \gamma_{23} \cos \beta, \tag{2.2}$$

$$\gamma_{12} \sin \alpha = \gamma_{23} \sin \beta. \tag{2.3}$$

For the case of the phases 1, 2 and 3 being two fluids and a solid the situation that is normally observed is depicted in fig. 2.2(b). This is not strictly an equilibrium situation; in fact, it is clear that (2.3) cannot be obeyed and we must suppose that at the point of contact of the three interfaces a local 'puckering' of the solid surface occurs, possibly like that shown in the magnified diagram in fig. 2.2(c). In this way the system establishes equilibrium in the locality of the intersection, but the solid

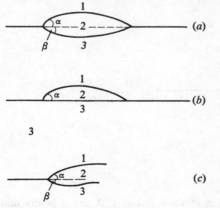

Fig. 2.2. Schematic diagram of a 'sessile drop'.

surface some distance away will be very slowly trying to achieve the situation represented in fig. 2.2(*a*). The normal form of (2.2) for a vapour–liquid–solid situation, for instance

$$\gamma_{SV} - \gamma_{SL} = \gamma_{LV} \cos \alpha \qquad (2.4)$$

which is represented in fig. 2.2(*b*) is therefore not strictly valid, but is usually a good approximation due to the small value of β at equilibrium.

Another example of the method of interface intersections is given by the study of the equilibrium shape of the grooves formed where a solid surface intersects a grain boundary. However, in applying this method to solid surfaces the simple formula (2.1) no longer applies due to the anisotropy of γ. This is because the surface tensions are acting in such a way as to change the orientation of the other interfaces, which in turn changes the surface tension of these interfaces. This orientation effect may be described by forces acting on each boundary tending to orientate it to a low γ orientation. The problem has been considered in detail by Herring (1951*b*), and, indeed, the extra terms needed in (2.1) to describe the equilibrium of solid interfaces are now normally referred to as the 'Herring torque terms'. To understand these terms consider a system of interfaces as in fig. 2.1, but with the γs now relating to solid surfaces and therefore being functions of orientation. Now let the intersection of the interfaces suffer a virtual displacement in the plane of the interface 2–3 by an infinitesimal amount from the position 0 to some new position P. We suppose that in this process the interface 1–2 becomes kinked at A and similarly the interface 3–1 becomes kinked at B as shown in fig. 2.3. AP and $BP \gg OP$ but are still infinitesimal. For unit length of interfaces perpendicular to the plane of the

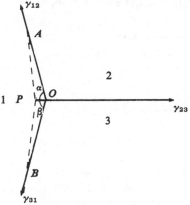

Fig. 2.3. Intersection of three phases 1, 2 and 3 before and after a virtual displacement of the intersection boundary.

diagram we may evaluate the change in free energy. Neglecting, for the time being, the effects of orientation change in 1–2 and 3–1 the free energy change is

$$dG_1 = (\gamma_{23} - \gamma_{12} \cos \alpha - \gamma_{31} \cos \beta) \cdot \overline{OP}. \qquad (2.5)$$

In addition, there are terms due to the changes of surface tensions caused by the local changes in orientation of the interfaces, and these are given by

$$dG_2 = \frac{(\partial \gamma_{12})}{\partial \alpha} \delta \alpha \cdot \overline{AP} \qquad (2.6)$$

and

$$dG_3 = \frac{(\partial \gamma_{31})}{\partial \beta} \delta \beta \cdot \overline{BP} \qquad (2.7)$$

but

$$\sin (\delta \alpha) = \overline{OP} \cdot \frac{\sin \alpha}{\overline{AP}} \simeq \delta \alpha$$

for small change in α, and similarly

$$\delta \beta \simeq \overline{OP} \cdot \frac{\sin \beta}{\overline{BP}}.$$

Thus the total free energy change associated with the virtual displacement is

$$dG = \overline{OP} \cdot \left\{ (\gamma_{23} - \gamma_{12} \cos \alpha - \gamma_{31} \cos \beta) \right.$$
$$\left. + \sin \alpha \frac{\partial \gamma_{12}}{\partial \alpha} + \sin \beta \frac{\partial \gamma_{31}}{\partial \beta} \right\}. \qquad (2.8)$$

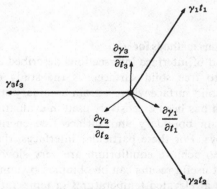

Fig. 2.4. Forces acting at an intersection boundary of three phases showing both the normal surface tension and the 'Herring torque terms'.

At equilibrium, this change should be zero so that

$$\gamma_{23} - \gamma_{12}\cos\alpha + \sin\alpha\frac{\partial\gamma_{12}}{\partial\alpha} - \gamma_{31}\cos\beta + \sin\beta\frac{\partial\gamma_{31}}{\partial\beta} = 0. \quad (2.9)$$

To solve the problem in its most general form one must, of course, allow for virtual displacements in directions other than the special one considered in the example of fig. 2.3 (here we have restricted the displacement OP to lie in the plane of one of the interfaces). It should be clear from the above analysis, however, that the equilibrium configuration of three interfaces, having surface tensions γ_1, γ_2 and γ_3 acting in directions t_1, t_2 and t_3 from a common intersection is given by

$$\sum_{i=1}^{3}\left(\gamma_i t_i + \frac{\partial\gamma_i}{\partial t_i}\right) = 0. \quad (2.10)$$

Thus the equilibrium may be depicted by the interaction of the three surface tensions terms $\gamma_i t_i$ acting tangential to each interface and three additional forces, the 'Herring torque terms', $\partial\gamma_i/\partial t_i$ which act normal to each interface. In the simple case where the terms γ_i are independent of orientation the torque terms $\partial\gamma_i/\partial t_i$ are zero and (2.10) reduces to the simple form described by (2.1).

One further point worthy of mention is the case where one of the interfaces corresponds to a cusp in the γ-plot for that interface. Clearly, in this case $\partial\gamma_i/\partial t_i$ is indeterminate. Consequently, as Herring (1951b) has pointed out, (2.10) can be satisfied by any value of $\partial\gamma/\partial t$ lying between the values on the two sides of the cusp; such a boundary will have the form of a plane facet.

2.2   Interface intersections methods for γ_{SL}

One example of the method of interface intersections described above, which has been applied to free solid surfaces, is the study of the equilibrium structure of a solid surface where a grain boundary emerges at the surface. This method has been extensively used in evaluating the relative magnitudes of grain boundary and surface free energies in both pure metals and alloys. For these particular interfaces, the processes that must operate to achieve equilibrium are very slow being controlled by diffusion, but useful results can be obtained by annealing specimens for several days at elevated temperatures (a temperature of the order of four-fifths of the melting point is typically chosen). At first glance the method would appear to be ideal also for a comparison of grain boundary and solid–liquid interfacial energies, but in general two major difficulties are encountered.

Firstly, there is the experimental difficulty of devising a system of observing the equilibrium shape of such a boundary. Most materials of interest technologically (metals in particular) are opaque and so the equilibrium cannot be observed with the solid and liquid phases co-existing. Observations must therefore be made by one of the two techniques used for observing the shape of the solid–liquid interface during solidification. The first of these is to snatch the solid from the liquid and observe the decanted interface. Unfortunately, in the decanting operation a thin liquid film always adheres to the surface which quickly freezes and thereby obliterates the detailed structure of the surface. The second method is to quench the melt suddenly, and then section and etch the resulting material to observe the sudden transformation of microstructure and segregation introduced by the quench beyond the then 'frozen-in' interface. This method is only applicable to alloy systems for which some compositional discontinuity can be observed at the quenched interface after treating in such a way as to show up variations in composition (e.g. etching). It is therefore impossible to apply the technique to pure metals.

The second, more fundamental difficulty with experiments of this kind is that the free energy of a grain boundary having angular misorientations more than 15° or 20° is greater than twice the solid–liquid interfacial free energy. This means that an equilibrium shape for the intersection with the solid–liquid interface is no longer possible and the liquid penetrates down the grain boundary rather than forming a cusp-like depression of definite angle at the intersection.

These difficulties have been overcome by Glicksman & Vold (1969) in their experiments to measure the solid–liquid interfacial free energy of bismuth. The first problem, that is of observing the interface, was overcome by using transmission electron microscopy of thin films of the metal, which therefore permitted direct observation of the two phases in equilibrium, and the second difficulty was overcome by studying the intersection of the solid–liquid interface only with low angle tilt grain boundaries of free energy less than $2\gamma_{SL}$.

The specimen was in the form of a thin (about 2000 Å) vapour-deposited film of bismuth sandwiched between thin carbon films, mounted on the hot stage of an electron microscope. In this way the temperature was raised to near the melting point by the hot stage and the specimen was finally melted in the area being viewed by the heating effect of the observing electron beam. A radial temperature gradient with the highest temperature at the centre was therefore imposed on the specimen just in the region of the field of view.† Evaluations of the free surface energy of the solid–liquid interface were made by measuring the angles of the cusp-like depressions formed on the edge of the central liquid pool where symmetrical, low angle, pure tilt grain boundaries intersected the solid–liquid interface. The situation is shown schematically in fig. 2.5. By choosing symmetrical tilt boundaries torque terms

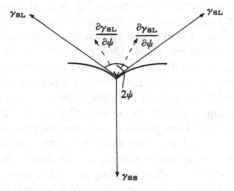

Fig. 2.5. Forces acting in the dihedral groove formed at the intersection of a symmetrical tilt boundary with the solid–liquid interface.

for the solid–solid interface are eliminated, and the equilibrium may therefore be represented by

$$\gamma_{SS} - 2\gamma_{SL} \cos \psi - 2 \frac{\partial \gamma_{SL}}{\partial \psi} \sin \psi = 0. \qquad (2.11)$$

In the experiments two parameters were measured, the dihedral angle of the groove in the solid–liquid interface, 2ψ, and the angle of tilt of the solid–solid boundary θ. Absolute values for the solid–liquid interfacial free energy were then deduced by calculating γ_{SS} from the known free energy of edge dislocations in the solid, assuming that the tilt boundary is composed of such dislocations whose density was proportional to the angle of tilt. Indeed, at the lowest angles of tilt (less than 1°) the angle and free energy could be deduced extremely accurately by counting the density of dislocations individually in the array

---

† The effect of this temperature gradient, other than as a stabilising influence was neglected. While the gradient in this case may be sufficiently small for this to be valid, temperature gradients generally can seriously distort a dihedral cusp at an interface intersection. Indeed, a recent paper (Jones & Chadwick, 1970) reports experiments in which this effect was utilised to measure γ_{SL} in certain materials by studying the shape of grooves at large angle grain boundaries where the dihedral angle was increased above its normal value of zero by the imposition of a suitable temperature gradient.

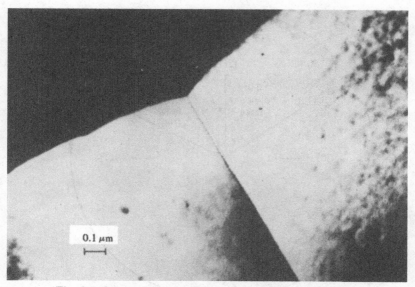

Fig. 2.6. Intersection of a low angle grain boundary with the solid–liquid interface in bismuth, after Glicksman & Vold (1969).

which composed the boundary. (An example of such a situation is shown in fig. 2.6.) Of course, uncertainties in the value of the energy of dislocations lead to uncertainties in the value of γ_{GB} and hence of γ_{SL}. This is an important limitation of the method. All the specimens took the form of thin basal plane platelets after annealing and so values of γ_{SL} were obtained only for surfaces perpendicular to the basal plane. Moreover, in the experiments of Glicksman and Vold, the values of γ_{SL} were obtained by neglecting the torque terms ($\partial\gamma_{SL}/\partial\psi$). The justification for this appears to be primarily that the results obtained in this way are consistent with this approximation.

The results obtained by Glicksman and Vold are shown in fig. 2.7 in the form of a partial γ-plot in the basal plane of bismuth. This shows that over a significant range of orientations γ_{SL} is approximately constant at a value of 61.3×10^{-3} J m^{-2}. This result will be discussed in relation to other values of γ_{SL} in a later section. In principle there should be a considerable degree of confidence in Glicksman and Vold's method compared with other less direct methods. Unfortunately though, conceptually so simple a method has not, as yet, been applied to any other materials. It seems likely to be considerably restricted in application because many materials rapidly evaporate or globulate at elevated temperatures especially under the conditions of high vacuum prevailing in the electron microscope.

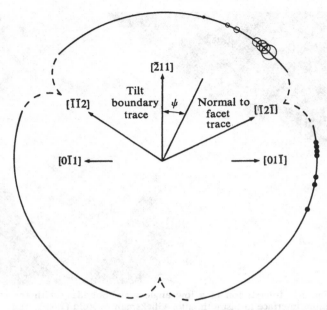

Fig. 2.7. Partial γ_{SL}-plot bismuth after Glicksman & Vold (1969). The dashed minima are simply inferred from the observation of facets at these orientations which may be either growth or equilibrium forms; if they are the latter then the minima will be as shown.

The alternative method of applying grain boundary grooving to determining solid–liquid interfacial free energies in pure materials involves a rather indirect approach through the study of alloys. It has already been mentioned that the grain boundary groove formed at the solid–liquid interface of a pure solid–melt system usually has a dihedral angle of zero. Thus if the grain boundary interfacial free energy is γ_{GB} this simply tells us that

$$2\gamma_{SL}/\gamma_{GB} < 1.$$

This relation does not always apply to alloys. Very often for alloys there is a non-zero dihedral angle (probably due to segregation effects at the grain boundary) so that γ_{SL}/γ_{GB} for the alloy system can be measured by simple annealing and quenching experiments, followed by metallographic analysis of the quenched interface at grain boundary grooves. Moreover, the equilibrium between solid and liquid in a bulk specimen is relatively easily attained in an alloy system because a small change in temperature results only in a slight shift of the position of the interface and the equilibrium compositions of the solid and liquid. For a pure material, such a change in temperature would result in total solidification or melting of the specimen (though in practice the latent heat absorbed or generated would buffer this action to some extent). It is possible, by studying the change in γ_{SL}/γ_{GB} for an alloy system as a function of alloy composition, to extrapolate the results to a pure material which is one of the components of the alloy. This is particularly likely to be useful in the case of an alloy having associated very low solid solubilities; in this case the grain boundary structure and energy may be assumed to remain constant throughout the alloy range, and thus extrapolation of results to the pure metal is more likely to be valid. Clearly, considerable caution is needed in applying extrapolation, but it does seem to have been applied with reasonable success to a few systems, notably by Miller & Chadwick (1967).

2.3   Homogeneous nucleation in the liquid–solid transition

While the methods of interface intersections outlined above provide a very direct means of determining γ_{SL}, and have been used extensively in measurements of solid–vapour, solid–solid, liquid–liquid and liquid–vapour interfacial free energies it has already been pointed out that it is generally difficult to adapt them to the study of pure material solid–melt interfaces. This is particularly true for the large number of materials which are of technological interest but are not transparent to visible light. One such group of materials is the metals; this is particularly disconcerting not only because they are technologically important, but also because they offer systems which should be relatively simple to treat theoretically. By far the largest number of measurements of the solid–liquid interfacial free energies in metals, and indeed in many other materials, have been made by the rather indirect method of determining the critical supercooling for homogeneous nucleation of the solid in the melt. In order to extract the value of the interfacial free energy from these experiments it is necessary to develop a theory of homogeneous nucleation.

We know that for the solid–liquid phase transition to occur the solid phase must have a lower free energy than the liquid at temperatures below the melting point, whereas the free energy of the liquid must be the lower above melting point. The melting point is, by definition the temperature at which the free energies of the solid and liquid phases are the same, so that the two phases can co-exist in equilibrium. The Gibbs free energies of the two phases are shown schematically in fig. 2.8. At the phase transition there is a discontinuity in gradient in the curve of G against T which represents the change in entropy on passing

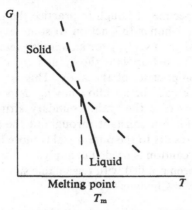

Fig. 2.8. Gibbs free energy of solid and liquid phases as a function of temperature near the equilibrium melting point T_m.

from one phase into the other. Thus, for the solid and liquid phases represented by the subscripts S and L at T_m we have

$$\left(\frac{\partial G_S}{\partial T}\right)_P - \left(\frac{\partial G_L}{\partial T}\right)_P = \frac{\Delta H_f}{T_m} \qquad (2.12)$$

where ΔH_f is the enthalpy of fusion (or latent 'heat' of fusion). On this basis it may be thought impossible to supercool a liquid (i.e. reduce its temperature below the melting point without the transformation to solid occurring), or to superheat a solid, as either of these produces a state of higher free energy than if the transformation had occurred. However, this conclusion neglects the effect of the free energy associated with the interface between the two phases. If the two faces are in equilibrium and a change of temperature is applied, then the transformation will advance in the case where the interfacial area and therefore the total interfacial free energy can remain constant in transforming further material, (as for example, in the advance of a planer interface along a bar of fixed cross-section). However, if a significant increase in interface area is involved in advancing the transformation, then the increased total interfacial free energy must be accounted for in determining whether an advance in the transformation does result in a reduction of the total free energy of the system. These considerations are particularly significant in considering the initial growth of a solid nucleus in a liquid below the melting point. In order for a liquid to transform to the solid phase at some temperature below the equilibrium melting point T_m by an amount ΔT, say, a small number of atoms or molecules in the liquid must come together to form an ordered array like the solid, which can then grow. As a simple model, we shall consider that the Gibbs free energy of a very small spherical nucleus is composed of a contribution from the interface between it and the liquid phase and one from its volume (which, having transformed, will have a lower free energy than the surrounding supercooled liquid). For a spherical nucleus, of radius r, the interfacial free energy term is simply $4\pi r^2 \gamma_{SL}$, where in this case γ_{SL} is some averaged value over all orientations which will depend in detail on the shape of the nucleus. Now, if we assume that $(\partial G_S/\partial T)_P$ and $(\partial G_L/\partial T)_P$ do not vary with temperature over the range of temperatures from $T_m - \Delta T$ to T_m, then the volume contribution to the free energy of the nucleus relative to the surrounding liquid phase is

$$-\tfrac{4}{3}\pi r^3 \frac{\Delta H_f}{T_m} \Delta T$$

i.e. the total excess free energy of the nucleus is

$$\Delta G = -\tfrac{4}{3}\pi r^3 \frac{\Delta H_f}{T_m} \Delta T + 4\pi r^2 \gamma_{SL}. \tag{2.13}$$

This is shown schematically in fig. 2.9 from which it should be clear that there is a maximum value of ΔG corresponding to some critical radius r^*. Thus there is a thermodynamic potential barrier to nucleation in creating a nucleus of radius r^*. Once a nucleus of radius greater than r^* has been created further transformation results in a continued

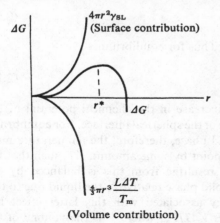

Fig. 2.9. The surface and volume contributions and the total excess free energy ΔG of a spherical solid nucleus within a supercooled liquid as a function of the nucleus radius r.

decrease in the free energy and so will proceed spontaneously. The value of r^* corresponds to the maximum value of ΔG and so corresponds to the condition $(\partial \Delta G/\partial r) = 0$. This gives, from (2.13),

$$r^* = \frac{2\gamma_{SL}T_m}{\Delta H_f \Delta T}. \tag{2.14}$$

It is useful here to introduce an alternative method of deriving the critical radius r^* which results from a consideration of the thermodynamic equilibrium across a curved interface. The simplest case to consider and one which may be readily applied to the spherical nucleus, is of a spherical interface of radius r. Suppose that the transformation of liquid to solid advances so that the sphere reaches an increased radius $r + dr$, where dr is very small. (This derivation is equally applicable to melting, or indeed any other transformation, but for convenience and brevity as well as applicability to the immediate situation, the transformation will be referred to here as solidification.) Now, if the chemical potential of the molecules in the solid and liquid phases are μ_S and μ_L and dn moles of material are transformed, then there will be a change in the Gibbs free energy of $(\mu_S - \mu_L)\,dn$. For equilibrium, this change in free energy must be equated with the increased free energy due to the formation of additional interfacial surface, $\gamma_{SL}\,dA$. For the sphere

$$dA = 4\pi\{(r + dr)^2 - r^2\} = 8\pi r\,dr \tag{2.15}$$

to first order in dr, and the change in volume

$$dV = v\,dn = 4\pi r^2\,dr, \tag{2.16}$$

where v is the molar volume. Thus for equilibrium

$$\mu_S - \mu_L = \frac{2\gamma_{SL}}{r}\,v. \tag{2.17}$$

This shows that there is an increase in the chemical potential of the solid phase due to the presence of the spherical interface. For equilibrium of a solid particle in the liquid phase, therefore, the temperature must be *below* the normal melting point by some amount, ΔT, such that the increased Gibbs free energy resulting from this is balanced by the decreased free energy of the solid phase relative to the liquid due to this supercooling. The free energy associated with this latter effect has already been shown to be $\Delta H_f V \Delta T/T_m$, where V is the volume of the particle; the former term is

$$n(\mu_S - \mu_L) = V(\mu_S - \mu_L)/v.$$

Thus, for equilibrium

$$\frac{\Delta H_{\mathrm{f}} \Delta T}{T_{\mathrm{m}}} = \frac{2\gamma_{\mathrm{SL}}}{r} \tag{2.18}$$

which gives, as before (cf. (2.14)),

$$\Delta T = \frac{2\gamma_{\mathrm{SL}} T_{\mathrm{m}}}{\Delta H_{\mathrm{f}} r}. \tag{2.19}$$

In fact this alternative argument is simply the differential form of the first approach including a reference to chemical potentials to gain familiarity with these quantities. However, it can also be applied to a general surface with radii of curvature r_1 and r_2 to give

$$\Delta T = \frac{\gamma_{\mathrm{SL}} T_{\mathrm{m}}}{\Delta H_{\mathrm{f}}} \left(\frac{1}{r_1} + \frac{1}{r_2} \right). \tag{2.20}$$

This is the form of the Gibbs–Thomson relation which is most useful for the application to the solid–liquid interface in solidification and melting as it expresses the effective change in melting point at a curved interface.

Returning to the homogeneous nucleation problem, we have now evaluated the radius of a spherical solid particle in equilibrium with liquid supercooled by some amount ΔT ((2.14) and (2.19)). However, as the first treatment shows particularly clearly, in the case of such a supercooled liquid the relation describes a metastable situation since a slight perturbation to a larger radius of curvature results in a spontaneous growth of the solid phase until such time as all the liquid is solidified, or the latent heat emitted brings the temperature of the system back to T_{m}. Thus, in the absence of some simpler method of nucleation, at impurities or surfaces which will be dealt with in the next section, a liquid can be supercooled until it reaches some temperature at which the size of critical nucleus corresponding to that supercooling is such that it is likely to be formed out of the random fluctuations of the molecules in the liquid. When a fluctuation occurs such that an array of molecules come together for a finite time in a solid-like array of dimensions equal to, or greater than, a sphere of radius $r*$ for that supercooling, then homogeneous nucleation will occur.

Now for such an array of i molecules we can apply Boltzmann statistics whence in a total of n molecules per unit volume, the number of arrays of i molecules on average is

$$m_i = n \exp\left(-\Delta G / kT\right) \tag{2.21}$$

per unit volume where k is Boltzmans constant and ΔG is the excess free energy of this array of molecules, as given by (2.13). Thus, the number of critical nuclei is given by say

$$m_i^* = n \exp (-\Delta G^*/kT) \tag{2.22}$$

where ΔG^* is the value of ΔG with r^* substituted for r. Now, in general, the nucleation rate is given by (for unit volume)

$$I = Dn^*m_i^* \tag{2.23}$$

where D is the diffusion or collision frequency of atoms or molecules per atom or molecule and n^* is the number of atoms or molecules on the surface of the critical nucleus. For a liquid, or solid, one may write approximately

$$D = \frac{kT}{h} \exp \frac{(-\Delta G_A)}{kT} \tag{2.24}$$

taking kT/h as a typical vibrational or 'jump' frequency and where h is Planck's constant and ΔG_A is the activation energy of the diffusion process. Thus

$$I = \frac{kT}{h} n^*n \exp \frac{(-\Delta G^* - \Delta G_A)}{kT}. \tag{2.25}$$

In fact this derivation assumes that the number of critical nuclei in equilibrium at some temperature is not influenced by the growth of any one of them (which will be very rapid when it occurs). This is not a valid assumption and a more rigorous treatment for the solid–liquid case by Turnbull & Fisher (1949) gives

$$I = \frac{nkT}{h} \exp \frac{(-\Delta G^* - \Delta G_A)}{kT}. \tag{2.26}$$

It is found that I changes very rapidly from an insignificant to a highly significant value over a very small temperature range. For this reason the observed nucleation temperature can be related to a quite well-defined temperature in the theory which is relatively insensitive to the pre-exponential factor (thus making the theoretical refinements of little significance). As an indication of the sharpness of the nucleation 'transition', Turnbull (1950a), has observed that the nucleation rate of droplets of liquid mercury changes by a factor of ten within a temperature interval of $1\frac{1}{2}$ deg at an undercooling of about 60 deg. Using the above theory, experimental nucleation results can thus lead to a value of γ_{SL} averaged (in some way as yet unspecified) over all orientations, determined from measurements of the maximum supercooling to which liquids can be subjected before homogeneous nucleation occurs.

Experimentally, the principal difficulty of the method lies in ensuring that truly homogeneous nucleation is observed. If a bulk liquid is cooled under normal conditions, undercooling of more than a few degrees are rarely observed, where the degree of supercooling that is theoretically possible before homogeneous nucleation occurs can be several hundred degrees. This is because impurities and the surfaces of the containing vessels can often act as heterogeneous nucleants at quite small supercoolings. The essence of the experimental technique is therefore to divide the liquid to be supercooled into small particles (typically 10 to 100 microns) in the hope that at least some of these will contain no heterogeneous nucleants and so will nucleate by a true homogeneous mechanism. Some of the first experiments were performed by subdividing the particles and keeping them separate by thin oxide films, or by suspension in some suitable fluid (Vonnegut, 1948; Turnbull, 1949) and measuring the nucleation rates and temperatures of the aggregate with a dilatometer. However, most of the later results on metals, which are the simplest materials to understand, have been achieved by observing the solidification temperatures of isolated particles on a microscope hot stage (Turnbull & Cech, 1950). The principal advantage of this technique is that it allows studies of high melting point metals to be made relatively easily. Moreover, there is some evidence that in some materials a surface oxide film on the liquid drops can act as a heterogeneous nucleant and so give spuriously small maximum supercoolings; in this technique such materials can be studied under an atmosphere of hydrogen to reduce the oxide film. In other cases, inert gas atmospheres are used. One problem in observing isolated particles under the microscope is that a substrate must be in contact with the specimen and may therefore act as a heterogeneous nucleant; this is overcome by using substrates of freshly blown quartz or pyrex glass which are unlikely to provide preferential sites for nucleation. Observations of the nucleation and solidification events are made by means of the change in the surface appearance associated with the solidification, or in some cases the 'blink' or sudden brightening of the particle when it solidifies. This is due to the sudden increase in temperature (recalescence) of the specimen caused by the generation of latent heat when rapid solidification follows the nucleation event. More recently other methods of detecting the transition such as by the change in a transmission electron diffraction pattern (Stowell, 1970) or even in a nuclear magnetic resonance (NMR) signal (E. F. W. Seymour, private communication) have been used. Results for the maximum supercooling, and the value of γ_{SL} derived from them, are listed in table 2.1 for a number of metals, and also for water. These are the values given by Turnbull (1950b) deduced largely from his, and his co-workers' experiments. Many more complex organic materials have also been studied in this way and a more complete list is given in the review paper by Jackson (1965).

Table 2.1 *Results of homogeneous nucleation experiments (after Turnbull, 1950b)*

Material	Melting point (°K)	Maximum observed undercooling (deg) (50 μm droplets)	Solid–liquid interfacial free energy ($\times 10^{-3}$ J m^{-2})
Mercury	234.3	58	24.4
Gallium	303	76	55.9
Tin	505.7	105	54.5
Bismuth	544	90	54.4
Lead	600.7	80	33.3
Antimony	903	135	101
Aluminium	931.7	130	93
Germanium	1231.7	227	181
Silver	1233.7	227	126
Gold	1336	230	132
Copper	1356	236	177
Manganese	1493	308	206
Nickel	1725	319	255
Cobalt	1763	330	234
Iron	1803	295	204
Palladium	1828	332	209
Platinum	2043	370	240
Water	273.2	39	32.1

In addition to calculating the particular form of the mean interfacial free energy given by the supercooling results, Turnbull also calculated the 'gram-atomic free surface energy', σ_g which he defines as the interfacial free energy of a surface, one atom thick, containing Avagadro's number of atoms. He compared this figure with the latent heat (enthalpy) of fusion, and the graph in fig. 2.10 shows this comparison. It shows that most of the metals, in particular the cubic metals together with mercury and tin, appear to fit into a class for which, numerically

$$\sigma_g \simeq 0.45\Delta H_f. \tag{2.27}$$

A small number of materials, namely water, bismuth, antimony and germanium fall into a separate class for which

$$\sigma_g \simeq 0.32\Delta H_f. \tag{2.28}$$

The existence of this distinction is not entirely surprising in that the second group of materials all have complex solid structures and expand

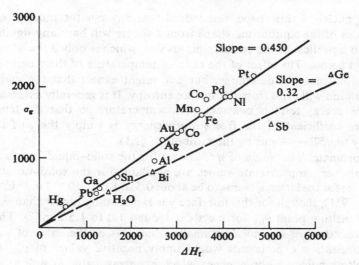

Fig. 2.10. Experimental values of the gram-atomic interfacial free energy σ_g versus latent heat of fusion (after Turnbull, 1950b).

on solidifying; on the other hand, gallium also has these properties but appears to behave in the manner of the first class. Unfortunately, in the absence of a detailed theory it is not possible to draw any definite conclusions from these observations. For instance, it is noticeable that σ_g for aluminium is rather low and while this may well be a result of oxide film effects, which are particularly pronounced in this material, it indicates that similar spurious effects may be responsible for the apparently low values for the second class of materials. There does, however, seem to be some hope that other simple cubic metals might fall into the first group so that their interfacial free energies may be approximately predicted on the basis of (2.27).

While the homogeneous nucleation experiments undoubtedly provide the greatest wealth of information on the values of γ_{SL}, the results should nevertheless be treated with some caution. In particular the values obtained correspond to some, as yet unknown, average for γ_{SL} taken over all orientations of the nucleus and furthermore they actually correspond to the values for very small particles at temperatures markedly removed from the melting point (the value of γ_{SL} close to the melting point is usually of greatest importance). It has been suggested that the average value of γ_{SL} will correspond to that for a nucleus of equilibrium shape, but while such a shape would correspond to the smallest thermodynamic potential barrier for nucleation, it is not clear that random collisions forming the nucleus cluster will always generate critical nuclei of this shape, nor indeed that any but the most gross variations of the equilibrium shape from a sphere will have any significance in a particle of the critical nucleus size, which is only a few atom spacings across. The effect of the reduced temperature of the measurement is not easy to determine, but one might expect that the main contribution will arise from the surface entropy. It is generally assumed that the energy term is constant with temperature so that the temperature coefficient of the free surface energy is simply the surface entropy ($\partial \gamma / \partial T = -\eta$ using the symbols of §1.1).

Unfortunately, no value of η is known for the solid–liquid interface and only very approximate values are available for the solid–vapour interface; for the latter, η seems to be about 0.5 to 1.0 $\times$ 10^{-3} J m^{-2} C^{-1} (Udin, 1951), though for this interface γ is significantly higher than γ_{SL} (at the melting point γ_{SV} for metals is around 1.0 to 1.5 J m^{-2}). The absolute values of η_{SL} must be smaller than this or the value of γ_{SL} from nucleation experiments would imply negative values of γ_{SL} at the melting point – a consequence which is unreasonable as well as at variance with the few experimental values obtained near the melting point! If we suppose that η_{SL} is proportionally smaller than η_{SV} by the approximate ratio (γ_{SL}/γ_{SV}), this would imply that the values of γ_{SL} from nucleation experiments are perhaps 5 or 10 per cent higher than the value at the melting point. It is interesting to note that this would *increase* the discrepancy between the values of γ_{SL} for bismuth measured by nucleation experiments and by grain boundary grooving observations. From homogeneous nucleation at 90 degrees below the melting point it is found that $\gamma_{SL} = 54.4$ erg cm^{-2}, whereas from the grain boundary grooving method at the melting point $\gamma_{SL} = 61.3$ erg cm^{-2} perpendicular to the basal plane. The increased discrepancy could be a result of nucleus shape effects but two alternative explanations deserve consideration.

The first of these is a result of recent considerations of nucleation theory by Lothe & Pound (1962, 1968). The details of their analysis are beyond the scope of this book, but by considering the effects of translation and rotation of individual nuclei they suggest that the value of γ_{SL} deduced from earlier theories may be low by factors of the same order as the discrepancy of the bismuth results. These ideas have been criticised by other authors, however, and it seems possible that further work may yet be needed to resolve the arguments. The second point relates to the thickness of the solid–liquid interface; in homogeneous nucleation theory it is assumed that the interface is well defined, or at least that its thickness is small compared with the size of the critical nucleus. However, various considerations (see chapter 4, and Hilliard

& Cahn (1958)) indicate that the solid–liquid interface may often be several atom spacings thick – i.e. of the same order as the critical nucleus size. If this is the case, then clearly some correction must be applied to the homogeneous nucleation results and, in the words of Hilliard & Cahn (1958) that γ_{SL} determined from the supercooling data 'will be too low by a presently unknown factor'.

Finally, some recent experiments have suggested that much lower undercoolings are possible than those observed by Turnbull, making his values for γ_{SL} as much as a factor of 2 in error (e.g. Stowell (1970) who also includes references to other recent work on nucleation).

2.4   Heterogeneous nucleation in the liquid–solid transition

In the previous section it has been stressed that experimental determination of γ_{SL} from nucleation experiments are only meaningful (i.e. can only be interpreted simply), when the nucleation is truly homogeneous. While it is of relatively little interest in terms of the determination of γ_{SL} it seems useful to explain briefly the principles involved in the nucleation of the solid phase on some particular solid material in contact with the melt, generally known as heterogeneous nucleation.

The necessary condition for a heterogeneous nucleant to be effective is that the interfacial free energy of the nucleant–solid nucleus interface should be less than the sum of the interfacial free energies of the nucleant–liquid and the liquid–solid interfaces. This is the condition for a particle of the solid nucleus material to wet (i.e. form a finite contact angle with) the nucleant, in the presence of the liquid phase of the nucleating material. This type of equilibrium situation is shown in fig. 2.2(*b*) where phase 1 is the liquid, phase 2 the solid (nucleus) and phase 3 the nucleant. The effect of this situation is to permit a nucleus presenting a radius of curvature equal to the critical radius for homogeneous nucleation to be formed from a far smaller number of atoms than that in a free nucleus floating in the melt. Thus the amount of undercooling required for this event to occur spontaneously will be much less than that required for homogeneous nucleation. This is one reason, therefore, why nucleation experiments may appear to give anomalously low undercoolings. The nucleant may be in the form of undissolved impurity in the melt or it may be the surface of the container or substrate of the cooled liquid.

It is important to note, however, that in general the interfacial free energy of a solid–solid interface will be *higher* than the sum of the two interfaces present when a thin 'sandwich' of the melt of one of the solids is introduced between them. The essential reason for this is that at a solid–solid interface there will be a marked mismatch of atomic spacings, resulting in strain in the solids near the interface, and thus energy is stored in the strain fields in the solids: introducing a liquid between the solids relaxes these strains, and, if the liquid is a melt of one of the solids, does not result in a marked increase in free energy due to the chemical discontinuity at the interface. Thus a heterogeneous nucleant will only be active for a nucleus when certain conditions on both the crystallographic orientations of nucleant and nucleus surfaces, and their mutual orientation, are satisfied such that the degree of atomic mismatch is low, and therefore the strain energy is low. It is for this reason that Turnbull used freshly blown quartz glass substrates for his homogeneous nucleation experiments. The non-crystallinity of the quartz makes suitable matching of atomic planes impossible, and with freshly blown quartz there is little chance of foreign particles being present.

Some attempts have been made to study heterogeneous nucleation under controlled conditions but the work involves many experimental difficulties and often gives erratic results. The results do seem to show, however, that in nucleus–nucleant systems there are usually several pairs of preferred relative orientations which can be active, and that undercoolings can be as low as a few degrees rather than the tens or even hundreds of degrees associated with homogeneous nucleation.

In addition to true heterogeneous nucleation other factors of less fundamental importance can also introduce serious errors into the measurement of undercoolings. One example of this is the trapping of small particles of the solid phase of the material under study in fine crevices in the container walls. Under suitable geometrical conditions of crevice a particle of the solid will present a concave interface to its melt and will therefore be in equilibrium *above* the melting point. On lowering the temperature below the melting point the solid will immediately grow into the melt from this site indicating negligibly small undercooling. All of these sites should be 'melted out', however, by adequate heating of the melt prior to the nucleation experiment.

2.5   Nucleation in the solid–liquid transition

On the basis of the theory of homogeneous nucleation presented in §2.3 it might seem reasonable to suppose that the theory could apply with equal validity to the solid–liquid transition. A reversal of sign in the undercooling (to superheating) and in the radius of curvature (to give a spherical liquid nucleus) would predict the same critical radius for homogeneous nucleation of melting. One might further deduce that the degree of superheating required for homogeneous nucleation of melting would be about the same magnitude as the undercoolings required in the homogeneous liquid–solid transformation. Of course some account must be taken of the inherent stability of positions of atoms in the solid which might be expected to increase the superheating required for the initial nucleus of liquid to be formed. Experimentally, however, superheatings are very rarely observed and in general a solid melts as soon as it reaches its melting point. In fact various authors have from time to time put forward the thesis that superheating a solid is *impossible*. This seems to be an unnecessarily extreme view and certainly one that is not always borne out by experiment. The general absence of any observable superheating appears to be a consequence of the relative magnitudes of the free energies of free solid, free liquid and solid–liquid interfaces in a solid–melt system. For most materials it is possible to write

$$\gamma_S > \gamma_L + \gamma_{SL}. \tag{2.29}$$

This means that if a free solid surface is at the bulk melting point temperature of that solid, then it is energetically favourable for a thin molten film to form on that surface. Moreover, assuming that the surface is planar, at least relative to the degree of the curvature effective in capillarity (see (2.19)) then there is no nucleation barrier to prevent this process occurring spontaneously; that is, there is no intermediate state of higher total free energy than the all solid and solid plus molten film stage. Indeed, the inequality (2.29) tells us that it is favourable for surface melting to occur *below* the equilibrium melting point for the bulk material; the small gain in free energy due to the presence of liquid phase in the film below the melting point will be more than balanced by the reduction in free energy achieved by relieving the solid of some of its free surface energy. However, rough calculations show that this effect will not be significant until the temperature is very close to (very much less than 1 degree from) the bulk melting point (see chapter 4). Nevertheless, the inequality (2.29) does mean that when a solid is raised to its bulk melting point, the free surface will spontaneously melt in this way, so that no superheating of the solid will be possible. This evidently accounts for the lack of observations of superheating. However, for some materials (2.29) does not appear to be true and hence they can be externally superheated (see §8.6). For 'normal' materials there remain several possibilities for achieving superheating experimentally. Essentially there are two means of achieving this. Firstly, the solid may be heated internally so that the surface is never

Fig. 2.11. Tyndall stars in an ice crystal, after Nakaya (1956).

allowed to reach the melting point while the interior is heated further, and secondly there is the possibility of preventing the outer surface from melting by reducing its free energy in some way. One way of achieving this latter effect would be to have the solid enclosed in another solid forming low energy solid–solid boundaries all round – such an enclosing solid would have to be a heterogeneous nucleant for the liquid phase of the material to be superheated (see §2.4). An example of this situation is the case where the solid superheated in crevices as mentioned in the previous section. However, the most hopeful method for a controlled experimental investigation seems to be that of internal heating. This method has produced a few fairly clear examples of superheating.

The best known of these experiments are those on ice which can be internally heated by focussing infra-red radiation on to a point within the solid. These experiments were first performed by Tyndall in 1858 who focussed sunlight on to ice crystals in Alpine glaciers, although the effect has been studied by many workers in more recent years. Tyndall noticed that when ice was heated in this way the liquid phase nucleated and grew within the solid in shapes resembling ice dendrites growing in water; these 'negative crystals' have since become known as Tyndall Stars and an example is shown in fig. 2.11. The dendritic form is a clear indication that superheating has taken place; as will be seen later solid dendrites are a result of interfacial instability due to super-cooling in the liquid from which they are growing. More recently workers have estimated the amount of superheating for nucleation of these stars from their experiments; the values so obtained generally seem to be less than 1 degree.

There have been other experiments in addition to those on ice. In 1939 two Russian workers Kaykin & Bené performed experiments on the superheating of tin. They heated rod-shaped specimens by passing high electrical currents through them so that the centre of the material would be hotter than the surface, due to heat losses from the surface. To further exaggerate this effect the outer surfaces were cooled by blowing air across them. For polycrystalline specimens melting always started inside the rod (where the temperature is highest) and the liquid metal burst through the surrounding solid. However, in the purer single crystal specimens, melting commenced at the outer surface, implying that the solid inside has been superheated to 1 or 2 degrees above the melting point without any nucleation of melting occurring. The internal melting of the polycrystalline samples is to be expected; it has already been mentioned that for grain boundaries the free energy is usually greater than twice the solid–liquid interfacial free energy (§2.2). Thus a grain boundary will melt, as does a free surface, at, or fractionally below, the bulk melting point without any nucleation barrier.

In addition to these experiments on the overall superheating of relatively simple solids, it has been shown that it is possible to superheat by more than 100 degrees certain solids having very viscous melts. In such a case the solid begins to melt from the free surface but the rate of melting is so slow that it is possible greatly to superheat the solid encased in its own melt. It does seem, therefore, that crystalline solids can be superheated by significant amounts under suitable conditions. However, it is unlikely that homogeneous nucleation of melting could ever be observed in a solid. Any homogeneous nucleation theory must, inherently, assume that the starting material is homogeneous apart from the local density fluctuations which permit nuclei to be formed in a 'random' fashion. It would seem to be applicable therefore only to an ideal perfectly ordered solid. A real solid necessarily contains point defects even in thermal equilibrium and the best prepared crystals usually contain dislocations. It may in addition contain planar defects such as twins or stacking faults but these may be relatively easy to avoid. A solid therefore contains built-in heterogeneities fixed in time and position. One might therefore deduce that nucleation of melting in a solid is necessarily heterogeneous. Presumably theories could be developed to estimate the superheating possible for nucleation at a particular form of dislocation or point defect, but it seems unlikely that any detailed and meaningful experiment could be developed to test this theory or to determine the magnitude of the solid–liquid interfacial free energy.

2.6   Other methods for determining γ_{SL}

In addition to the methods discussed in §§2.2 and 2.3 a few other methods for determining γ_{SL} have been applied in a limited number of cases. These are based on the concepts of interface intersections or capillarity which have already been discussed but it is instructive to outline two specific approaches that have been adopted.

The first of these is a capillarity technique whereby γ_{SL} is determined from the variation of melting point with interface curvature. The specimen takes the form of a thin wedge held between two flats as illustrated in fig. 2.12. The reduction in the melting point for a cylindrical interface has already been given (cf. (2.20))

$$\Delta T = \frac{\gamma_{SL} T_m}{\Delta H_f r}. \tag{2.30}$$

In this case r, the equilibrium curvature, may be deduced from the position of the equilibrium interface, assuming prior knowledge of the wedge angle. The flats would typically be glass and so the angle of contact of the solid–liquid interface with the flats should be zero due to the relative magnitudes of the solid–flat, liquid–flat and solid–liquid interfacial free energies (see §2.4).

Fig. 2.12. Wedge technique for measuring equilibrium melting temperature as a function of interface curvature.

The second approach is a variation on the sessile drop type of experiment. This has recently been used by Mutaftschiev & Zell (1968) to determine γ_{SL} for the (0001) interface in cadmium. In their experiments they studied the angle of contact of the liquid on an emerging (0001) solid–vapour surface of a solid crystal grown within a drop of the liquid. The method is rather restricted, however, in that it is only applicable to a solid surface which is not wetted by its own melt (i.e. a surface for which (2.9) does *not* apply) and, as has already been pointed out, this is not normally the case. For certain close-packed surfaces in anisotropically structured materials (especially the basal plane of hcp materials) non-wetting may occur, but even the evidence for this, including the experiments of Mutaftschiev and Zell, is still in dispute. However, the technique does show the great variety of ways in which interface intersections methods can be applied to different situations. One particularly important aspect of this particular experiment is that it was performed under ultra-high vacuum conditions. The free surface energy of interfaces, and the relative surface energy of different orientations, can be drastically changed by quite a small concentration of certain impurities on the interface; when free surfaces are involved it is therefore important that the surrounding atmosphere should not contaminate the surface, a condition which can only be fulfilled with a reasonable degree of certainty under ultra-high vacuum.

2.7   Experimental determinations of the shape of the γ-plot for γ_{SL}

The methods described above offer means of measuring the absolute values of γ_{SL} at some specific orientation of the interface, or, in the case of nucleation experiments, of measuring some average value of γ_{SL} over all orientations. In principle much of the γ-plot for the interface may be determined by individual measurements of γ_{SL} at different orientations; apart from experimental difficulties this is extremely slow and tedious, and, as each measurement is independent of the others, large errors in relative values of γ_{SL} for different orientations may result. It is better, therefore, to make simultaneous measurements of the relative magnitudes of γ_{SL} for different orientations if this is possible. The means of achieving this has already been pointed out in §1.6. If a small particle is allowed to achieve equilibrium with its surroundings then its equilibrium shape gives directly the relative values of γ for all orientations represented on the surface of the particle.

There are two alternative experimental approaches for free solid surfaces. In one of these, the shape of very small, equilibrated, particles is determined by electron microscopy of the particles themselves and of their 'shadows' produced by evaporation at a glancing angle of some suitable 'shadowing' material. Alternatively, the equilibrium shape of small voids or gas bubbles can be studied directly by transmission electron microscopy.

Approaches of this kind are far more difficult for the solid–liquid interface. It has already been pointed out that small heat fluctuations radically change the position of the equilibrium interface in a solid–melt system. For this reason it is impossible to equilibrate a mixture of solid particles in their own melt, or melt droplets in their own solid, quite apart from the impossibility of observing the equilibrium should it be achieved. Experiments of this type can only be performed on alloy systems where small temperature fluctuations disturb the equilibrium only slightly because of the stabilising influence of compositional changes associated with interface motion. Even in these cases experiments cannot be performed satisfactorily on solid particles in the liquid due to the mobility of the particles in this matrix and hence their tendency to agglomerate and to float or sink. However, experiments may be performed on liquid alloy droplets in equilibrium with their solid; a partial quenching of a liquid alloy will result in some drops of liquid being trapped within single crystals of solid (between growing dendrite arms for example) and these will form suitable zones for study. At the same time some of the liquid segregated to the grain boundaries will offer opportunities for measuring grain boundary dihedral angles for absolute measurements of γ_{SL} as described in §2.2.

While these results are of interest they are likely to be unrepresentative of the results for the pure solid–melt system of immediate interest. In this area the only experiments available at present seem to arise from the technique of electron microscopy used by Glicksman and Vold. Liquid zones equilibrated within a solid film and held fixed by radial temperature gradients would seem to offer a likely means of observing equilibrium shapes of enclosed solid–liquid interfaces. So far, however, no quantitative results have been obtained by this technique, although a faceted, partially equilibrated liquid zone of bismuth in solid bismuth has been observed by these authors. Further investigations using this approach would seem to be well worthwhile, as no other comparable method seems to be available. The alloy technique is a very useful one but the extrapolation of the results to pure systems seems very dubious; certainly there is good evidence that the shape of the interface during solidification can vary greatly on alloying in quite small quantities as will be seen in the next chapter.

Further reading

Discussions of all the methods of determining γ_{SL} experimentally are notably lacking in the literature and so it is necessary to concentrate on individual methods in further reading. Nucleation is the most thoroughly dealt with and is covered in many metallurgical texts and reviews; of these we might mention those of Turnbull & Hollomon (1951), and more recently Uhlmann & Chambers (1965) and Jackson (1965). A review of all methods applied to solid–vapour interfaces has been given by Geguzin & Ovsharenkov (1962).

3   The structure of the solid–liquid interface

3.1   Singular and non-singular interfaces

In order to understand how a solid crystallises from its melt various theories have been developed which consider the transfer of atoms or molecules from the liquid to the solid state. At least one of these theories does not discuss explicitly the structure of the interface but rather by-passes this problem with some general assumptions and concentrates on the kinetics of the processes. Models of this type will be discussed later in this book. Of immediate interest, however, are the other crystal growth theories which attempt to resolve the problem of describing the structure of the interface in order to examine its mode of advance into the liquid (or into the solid).

The reason for approaching the kinetic problem in this way can be appreciated by considering two extreme models of the interface. The first is to consider a low index surface which is atomically smooth (a singular interface in the Frank notation). Alternatively, we can consider a rough interface in which several atom layers are involved in the transition between phases and the atoms in each layer fairly randomly positioned (except that presumably an atom in the solid must necessarily have a nearest neighbour below it in the solid so that the interface will have an irregular 'mountain range structure' on an atomic scale). Now consider the problem of addition of further atoms on to these interfaces. In the case of an atomically smooth interface an isolated atom placed on such a surface will be very weakly bound due to the relatively small number of nearest neighbours in the solid. Indeed if we consider the extra interface free energy introduced (were this a meaningful concept for a single atom), the energy of the system would in fact be increased. Further atoms placed next to this one would be more tightly bound but would position themselves in these adjacent sites preferentially due to the greater number of 'solid atom' nearest neighbours. The problem is in fact one of two-dimensional nucleation in that there is a ledge energy which must be equated with the decrease in free energy of each atom transferred from liquid to solid. Moreover, unless the undercooling is very large there will not be a new nucleation event until all ledge sites have been filled and the surface is once again atomically smooth. In the case of a rough interface, however, there will always be a large number of ledge sites into which new atoms can be easily accommodated. We can see, therefore, how the kinetic behaviour, in our case solidification and melting, might be markedly influenced by the equilibrium structure of the interface. This is assuming, of course, that the interface can still achieve some sort of equilibrium when moving, at least on a local scale. This is effectively the case if the rate of motion of particles over the interface is significantly faster than the net rate of arrival of particles, so that the interface can keep rearranging itself to its equilibrium configuration. A more detailed discussion of these kinetic processes will be given in chapter 8.

3.2   Jackson's (1958) theory of the interface structure

This approach to the general problem of crystal growth has been adopted by several authors. The work of Burton, Cabrera and Frank mentioned in chapter 1 was concerned with it since the 'surface melting' temperature defines for a free solid surface a dividing line between the two extremes of interfacial structure discussed above. Their work is primarily concerned with a free surface of a solid in that it considers a solid surface to be in equilibrium with itself. A simplified approach aimed particularly at the solid–liquid interface was presented by K. A. Jackson in 1958. The theory presents an extremely simplified model of the solid–liquid interface but has the merit of being remarkably successful in its predictions. The principal simplifications are that it considers only a two-level model of the interface, takes account only of nearest neighbour bonds in the solid and is based on a simplified Bragg–Williams statistical model (rather than the Bethe approximation used by Burton, Cabrera and Frank).

An interface which is initially singular or atomically smooth is considered and atoms are added to this in a single layer until a complete monolayer has been formed; the surface is then once again atomically smooth, but has advanced by one atom spacing. Now consider the change in free energy associated with adding a certain number of atoms randomly. In general the excess free energy will be of the form

$$\Delta G = \Delta E - T\Delta S + P\Delta V \tag{3.1}$$

at constant pressure and with the usual meaning attached to the symbols. In our particular case this may be written as

$$\Delta G = -\Delta E_0 - \Delta E_1 + T\Delta S_0 - T\Delta S_1 - P\Delta V \tag{3.2}$$

where ΔE_0, ΔE_1, ΔS_0, ΔS_1 and ΔV are all defined as positive and where ΔE_0 is the change in internal energy associated with the atoms being attached to the surface,

ΔE_1 is the change in internal energy associated with the atoms on the surface due to the presence of other adatoms on the monolayer,

ΔS_0 is the change in entropy associated with the adatoms passing to the solid phase from the liquid (or vapour) phase,

ΔS_1 is the configurational entropy associated with the different possible sitings of the adatoms on the surface, and

ΔV is the change in volume due to the atoms being associated with the new phase; for the solid–liquid transition this will subsequently be approximated to zero.

The problem now is to evaluate these quantities so that ΔG can be minimised to indicate the form of the equilibrium state. Consider one adatom and its nearest neighbours in the 'solid state'. It will have η_0 nearest neighbours in the solid which were present before any atoms were added, and a maximum possible number of adatom nearest neighbours, η_1. Moreover, the symmetry of the crystal structure will be such that, if growth were to continue until the atom considered became situated within the bulk of the solid, it would gain a further η_0 nearest neighbours in the process. Thus, if in the bulk solid an atom has ν nearest neighbours then

$$\nu = 2\eta_0 + \eta_1. \tag{3.3}$$

Now, if we define L_0 as the change in internal energy associated with the transfer of one atom from bulk liquid to bulk solid (not the normal latent 'heat', L, which is enthalpy), then an atom transferred from bulk liquid to an isolated site on a singular surface will suffer a change in internal energy of $L_0(\eta_0/\nu)$. In addition the η_0 nearest neighbour atoms in the solid will each suffer a similar change in internal energy of L_0/ν, so that for N_A adatoms

$$\Delta E_0 = 2L_0\left(\frac{\eta_0}{\nu}\right)N_A. \tag{3.4}$$

Further, any adatom site will on average have N_A/N nearest neighbour adatom sites filled (where N is the number of atoms in a complete monolayer on the surface considered) so that

$$\Delta E_1 = L_0\left(\frac{\eta_1}{\nu}\right)\frac{N_A}{N}\cdot N_A. \tag{3.5}$$

Note that in this case, unlike the expression for ΔE_0, no factor of 2 appears because both halves of each bond energy are automatically included by summing over all adatoms as the bonds considered here are only between adatoms and do not involve the substrate. Also, we have very simply

$$\Delta S_0 = \left(\frac{L}{T_E}\right) N_A \tag{3.6}$$

where T_E is the equilibrium temperature for the phase change (at which temperature the change will be assumed to occur). The configurational entropy, ΔS_1, is given by $k \ln W$ where W is the number of ways of arranging the N_A atoms in the N sites. This is given by

$$W = \frac{N!}{N_A! (N - N_A)!} \tag{3.7}$$

and so, using Stirling's approximation

$$\Delta S_1 = kN \ln \left(\frac{N}{N - N_A}\right) + kN_A \ln \left(\frac{N - N_A}{N}\right). \tag{3.8}$$

Finally, if we first consider the solid–vapour transition assuming that the volume of the solid phase is negligible compared with that of the vapour, and that the vapour phase is an ideal gas

$$P\Delta V = N_A kT \tag{3.9}$$

and similarly, the true latent enthalpy L is

$$L = L_0 + kT. \tag{3.10}$$

Combining the equations (3.2) to (3.10) gives

$$\begin{aligned}
\frac{\Delta G}{NkT_E} &= -\left(\frac{L_0}{kT_E}\right)\left(\frac{N_A}{N}\right)\left[\left(\frac{N_A}{N}\right)\left(\frac{\eta_1}{\nu}\right) + \frac{2\eta_0}{\nu}\right] \\
&\quad + \left(\frac{T}{T_E}\right)\left(\frac{N_A}{N}\right)\left[\left(\frac{L_0}{kT_E}\right) + \frac{T}{T_E} - 1\right] \\
&\quad - \left(\frac{T}{T_E}\right) \ln \left(\frac{N}{N - N_A}\right) - \left(\frac{T}{T_E}\right)\left(\frac{N_A}{N}\right) \ln \left(\frac{N - N_A}{N_A}\right).
\end{aligned} \tag{3.11}$$

If we make the assumption that $T = T_E$ and write

$$\alpha = \frac{L_0}{kT_E} \cdot \frac{\eta_1}{\nu}, \tag{3.12}$$

(3.11) may be re-written as

$$\begin{aligned}
\frac{\Delta G}{NkT_E} &= \alpha N_A \left(\frac{N - N_A}{N^2}\right) - \ln \left(\frac{N}{N - N_A}\right) \\
&\quad - \frac{N_A}{N} \ln \left(\frac{N - N_A}{N_A}\right).
\end{aligned} \tag{3.13}$$

Now if we consider the solid–liquid transition, and assume to a first approximation that there is no volume change on freezing or melting, (3.9) will become $P\Delta V = 0$, and (3.10) simplifies to $L = L_0$. Substitution of these amended equations arrives at the same final result of (3.13). This is because the two terms due to the non-zero value of ΔV in the solid–vapour case cancel out of (3.11) when the condition $T = T_E$ is used.

Equation (3.13) may be displayed graphically as a plot of $\Delta G/NkT_E$ against N_A/N (between 0 and 1) for different values of the parameter α, as shown in fig. 3.1. The principal feature of interest is that there are essentially two different types of curves. Those corresponding to $\alpha \lesssim 2$ show a minimum of excess free energy at $N_A/N = 0.5$; that is, when the monolayer is half-completed and so the surface is completely 'rough'

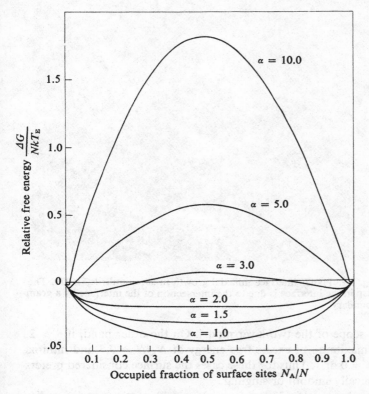

Fig. 3.1. Equation (3.13). Excess free energy versus monolayer occupation for various values of the parameter α.

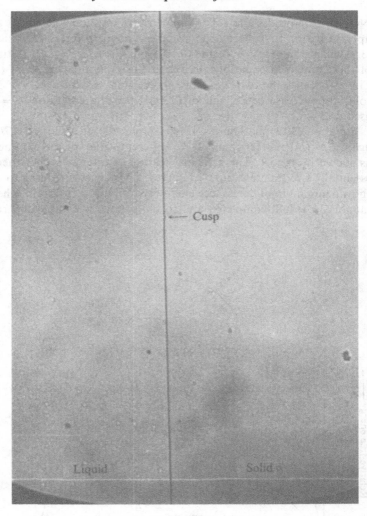

Fig. 3.2. Planar interface uniaxial growth in succinonitrile, × 270. The cusp in the interface is due to the intersection of the interface by a grain boundary.

within the scope of the two-layer model. On the other hand, if $\alpha > 2$, there is a maximum of excess free energy at $N_A/N = 0.5$ and minima near $N_A/N = 0$ or 1. Thus for these cases the surface considered prefers to be atomically smooth or singular.

The ideas discussed in §3.1 can now be used to make certain predictions about the behaviour of different materials in solidifying from their

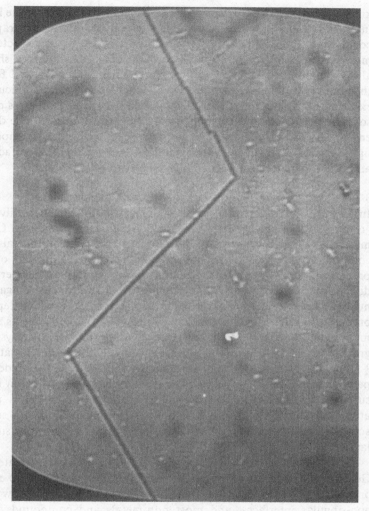

Fig. 3.3. Faceted growth of salol under conditions similar to those used in fig. 3.2, ×250.

melts. Materials for which $\alpha > 2$ for some surfaces will be expected to form singular solid–liquid interfaces and hence the growth form will display facets and the rate of growth will be controlled by the occurrence of steps on the interface (either created by two-dimensional nucleation or due to the presence of some crystalline imperfection). The different possible mechanisms involved in such faceted growth will be dealt with in greater detail in chapter 8; for the time being we shall refer to the kinetics of all faceted growth as 'source limited' implying that the limiting factor is the presence of suitable sites on the surface to trap the condensing atoms. Rather different properties are to be expected of materials for which $\alpha \leqslant 2$ for all surfaces, since the interface should always be non-singular and, according to Burton, Cabrera and Frank this should result in a growth process in which no nucleation difficulty is experienced. Therefore, the interface should follow the isotherms of the solid–liquid system and be rounded as opposed to faceted. The difference in the two types of behaviour can best be seen in the examples of figs. 3.2 and 3.3 which show uniaxial growth occurring on non-faceting and faceting materials under similar temperature gradients.

3.3   Comparison of Jackson's predictions with experiment
In general the values of α for most materials fall fairly readily into groupings which should determine the crystallisation properties. Low α materials ($\leqslant 2$) are principally the metals while most non-metals and compounds have much larger αs. In fact, α is really the product of two parameters, (L_0/kT_E) which is essentially the entropy of melting per mole divided by the gas constant R (multiplying top and bottom by Avagadro's number) and is a function of the material only, and (η_1/ν) which depends on the crystal structure and the surface considered. It will have its maximum value for the close-packed surfaces; for example, this crystallographic factor is $\frac{2}{3}$ for the (100) surface of a model simple cubic material, $\frac{1}{2}$ for (111) f.c.c. and also for (110) b.c.c. By the nature of the model this parameter can only be expected to make reasonable sense for low index surfaces, for which the initial singular state in the model is really atomically smooth, and for which a nearest neighbour model is still of some value in determining the binding of a surface atom in the surface layer.

Table 3.1 shows the values of (L_0/kT_E) for a number of materials. Generally, the success of the predictions of Jackson's analysis is extraordinarily good. Metals, the main low α-factor materials, grow with non-singular interfaces, and most non-metals and compounds grow with a faceted interface from their melt. One particularly useful feature of the work has been that it has led to the discovery that a small number of transparent organic compounds could be used as analogues of metals in studying solidification behaviour. Most organic compounds have large entropies of melting and grow in a faceted manner. There is, however, a very small group of such materials which have a very low entropy of melting (see table 3.1). The reason appears to be that they initially crystallise with their molecules randomly oriented on a b.c.c. or f.c.c. space lattice and thus have an entropy change associated with this

Table 3.1   *Values of L_0/kT_E for some materials. Those marked * are well known to be faceting on at least one orientation*

Material	L_0/kT_E	Material	L_0/kT_E
Potassium	0.825	Lead	0.935
Copper	1.14	Silver	1.14
Mercury	1.16	Cadmium	1.22
Zinc	1.26	Aluminium	1.36
Tin	1.64	Gallium*	2.18
Bismuth	2.36	Antimony	2.57
Germanium*	3.15	Silicon*	3.56
Water*	2.62	CBr$_4$	1.27
Cyclohexanol	0.69	Succinonitrile	1.40
Benzil*	6.3	Salol*	7

transition similar to that of a metal. At a rather lower temperature another phase transition occurs when the molecules are ordered and a further entropy change occurs. A study of these materials has shown that they do indeed behave as predicted by their low α-factors and solidify in the same way that metals do (Jackson & Hunt, 1965). Apart from the interest of these materials in providing a rather convincing confirmation of the value of Jackson's analysis, they also offer convenient transparent systems for the study of metallic solidification which is of considerable metallurgical significance. Fig. 3.2 shows one of these materials growing and may be compared with a normal (high α-factor) organic material such as that shown growing in fig. 3.3.

However, in addition to the main metal and non-metal grouping which appears to match the Jackson theory satisfactorily, there are a small number of materials falling between the two groups. Principally these are materials which in their general properties fall between the metallic and non-metallic classifications. These materials have values of $(\Delta S/R)$ lying between 2 and 3 and are not in strict order of faceting or non-faceting behaviour. Indeed, bismuth, for example, can display both types of behaviour. At low undercoolings so-called 'hopper' crystals are grown which have a faceted interface. At higher undercoolings dendritic growth, typical of metallic behaviour, is observed. These features are not surprising. Firstly, the value of α is also a function of the geometrical factor for the closest packed surface which is different for different structures and which is probably not a very good parameter for some of the rather complex structures found in many of these intermediate materials. Moreover, the transition from faceted to non-faceted growth of a borderline material as a function of undercooling can also be expected in a qualitative way, at least. The model outlined assumes zero undercooling; thus the totally empty monolayer has the same free energy as the full monolayer. For the case of finite undercooling, the curves in fig. 3.1 should be tilted down at the $N_A/N = 1$ end so that the first minimum (on curves for $\alpha > 2$) will move towards the 'rough' value of $N_A/N = \frac{1}{2}$. However, it is worth noting that the Jackson model predicts that for materials for which α is *less* than 2, growth is of the 'rough interface type' for all undercoolings; this is in direct disagreement with certain other models, as will be seen later. Unfortunately, there seems to be no decisive evidence experimentally on whether this is the case or not.

There is some evidence, however, that a small number of materials do behave other than as predicted by the α-factor theory. In particular, limited (and disputed) evidence exists that cadmium and zinc can grow from their melts with a faceted interface. Earlier work suggesting this also occurred in lead and tin has now been shown to be an artefact of the particular experimental technique used. In these experiments, the solid–liquid interface was studied by snatching the growing solid from its melt and observing the decanted interface. Subsequent work has shown that the facets observed on this surface were due to the freezing in air of the thin liquid layer which adheres to the decanted interface, and are not characteristic of the solid–liquid interface below.

One final case worthy of mention here is ice. As the table shows, $\Delta S/R$ is greater than 2 for this material, and in fact α is greater than 2 for the basal plane and less than 2 for surfaces normal to it. Thus, a typical growth form would be expected to be platelets having faceted slow-growing basal plane faces with rough interface growth on the edges of the plate. Such a growth form is indeed observed, although some of the many observations of growth in ice indicate that at least at very low growth rates, steps are observed on the interface independent of orientation, implying faceted growth behaviour everywhere.

3.4   Jackson's theory in perspective

Before leaving Jackson's theory and passing on to more complex models it is worthwhile putting the model and its implications more fully into perspective.

Firstly, it is interesting to compare this theory with the work of Burton, Cabrera and Frank mentioned earlier. The main difference is that in the theory of Burton, Cabrera and Frank the equilibrium considered is between a surface and the solid. In Jackson's analysis the presence of the second phase is also considered. Thus, while Burton *et al.* calculate that 'surface melting' should not actually be observed on low index metal surfaces because the bulk melting point is at a lower temperature, Jackson's theory says that the solid–liquid interface of such a material at the melting temperature should be rough (i.e. it should have 'surface melted'). However, if Jackson's theory is applied to the solid–vapour equilibrium (a more realistic comparison) then α for a metal is now a function of the entropy of vapourisation and is much greater than 2 (typically 10 or 14) leading to a singular interface as predicted by Burton *et al.* The difference is that whereas Jackson's analysis is restricted to a two-layer model of the surface and uses the simplified approach to the statistical mechanics of the problem (Bragg–Williams approximation†), Burton *et al.* consider models of the surface of varying degrees of complexity. They look at models having also 3 and 5 layers and use rather more refined statistical approximations (after Bethe †) and find that the exact form and position of the surface melting transition is a function of the model. All of these models, and also a simplified three-level Bragg–Williams theory due to Mullins (1959) investigate the surface melting transition relative to the value of a dimensionless parameter (kT/ϕ) where ϕ is proportional to the bond energy and hence to the binding energy of the solid for a free solid surface. Indeed, the essential difference between the simple form of the theories of Burton *et al.* and Jackson is primarily that Burton *et al.* are concerned with free solid surfaces and the variable of interest in the parameter (kT/ϕ) was temperature, whereas Jackson is concerned with the solid–liquid interface and so fixes $T = T_m$ and is interested in the variation of (T_m/ϕ) from one material to another. In addition the value of ϕ in Jackson's work is related to the latent heat of melting rather than vapourisation. A comparison between the various formal theories can

---

† In calculating the free energy of the interface we should strictly consider the energy of all the possible configurations of the interface. We can then construct a partition function $Z = \sum_i \omega_i \exp(-E_i/kT)$ where E_i is the internal energy of some configuration (in our model this is the sum of all broken bond energies for some configuration) and ω_i is the number of complexions of the interface having total energy E_i. The free energy is then given by $-kT \ln Z$. The solution of this problem (known as the Ising model) is complicated and can only be solved exactly for a simple two-dimensional model – i.e. a two-level model. To overcome this, various approximations can be made and these are discussed in detail in texts on statistical mechanics (e.g. T. L. Hill, *Statistical Mechanics*, McGraw-Hill, New York, 1956) or on order–disorder theory (e.g. E. W. Elcock, *Order–Disorder Phenomena*, Methuen, London, 1956). The simplest of these is the Bragg–Williams or 'zero order' approximation. This assumes that some average value of E_i can be substituted for all the E_i and that the value chosen should correspond to the most probable value (i.e. that corresponding to the largest value of ω_i). Taking this value as E, the partition function can then be written as $\exp(-E/kT) \sum_i \omega_i$ and so the free energy becomes $E - kT \ln W$ where $W = \sum_i \omega_i$. This approximation has therefore been explicitly made in Jackson's theory.

therefore be made simply by looking at the variation of some roughness parameter with (kT/ϕ). For this purpose we use the surface roughness parameters of Burton *et al.* defined as the excess internal energy of the equilibrium surface over that of an atomically smooth interface expressed in units of the latter (i.e. $(E - E_0)/E_0$). This is therefore proportional to the number of 'dangling' or solid–liquid bonds. For Jackson's model,

$$s = 4 \frac{N'_A}{N} \left(1 - \frac{N'_A}{N}\right)$$

where N'_A corresponds to the value for a minimum on the curve of $\Delta G/NkT_E$ versus N_A/N. A comparison of the results of various models is shown in fig. 3.4 for the particularly simple case of a (100) surface of a simple cubic material which constitutes the model used in the other calculations. s is plotted here against $1/\alpha$, the reciprocal of the Jackson α-factor. This shows that the more refined models, particularly that of Burton *et al.*, give a more smoothed-out transition, and that the steep part of the curves correspond to values of α rather greater than 2. However, within the framework of the assumptions used to apply the Jackson theory to real situations, these corrections are probably not

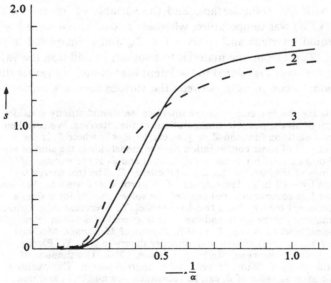

Fig. 3.4. Surface roughness s versus $1/\alpha$ for three different models and associated statistical approximations.
1. 3-level Bragg–Williams (Mullins)
2. 3-level Bethe (Burton, Cabrera and Frank)
3. 2-level Bragg–Williams (Jackson)

very meaningful. The smoothing out was in fact shown by Burton *et al.* to be a function of both the statistical treatment and the model used. The Bethe treatment applied to a two-level model also produces a very sharp transition, but this is not a feature of the exact treatment of the two-level model by Onsager (1944; and Onsager & Kaufmann, 1946).

Finally, it is interesting to note the form of the implications of the Jackson theory. As has already been pointed out, the model is essentially an equilibrium one. On the other hand it has been applied to experimental conditions where true equilibrium is definitely not established. In fact, the theory tells us very little which could be investigated experimentally about the equilibrium interface. It is known that, in general, when a surface has 'surface melted' it ceases to be singular, both in the normal meaning in this context of ceasing to be atomically smooth and in the mathematical context that the sharp cusp in the γ-plot corresponding to this surface will disappear. However, as the model makes no attempt to evaluate the relative free energy of different *orientations* of surface, the extent of this loss of singularity in the γ-plot is not known. Thus, the implication regarding equilibrium facets, for example, are very slight indeed; the equilibrium surface near the orientation of a non-singular minimum may be indistinguishable from a facet just as the facet due to a very shallow singularity cusp may not be large enough to be discernible as such. However, if the relationship between surface singularity and growth mechanism is as simple as has been suggested, the implications on growth forms and the presence of facets on the growing interface are far more definite. If singular interfaces exist they will be source limited in growth, and hence all other orientations will grow out rapidly simply leaving these slow-growing surfaces exposed. On the other hand, if the growth *mechanism* is the same on all orientations of surfaces, the anisotropies of growth should produce much less obvious effects on the growth morphology. Finally, however, it might be noted that there are reasons to believe that the true situation will not be as clear cut as this. The more refined theories particularly show that the singular to non-singular transition is not so sharply defined as is, for instance, the bulk melting transition. It is therefore not clear at what point the transition in behaviour should occur and it seems almost certain that there should be some cases where the results are not clearly defined on one side or other of the transition.

3.5   Temkin's *n*-layer model of the interface
In all the treatments of the surface structure so far considered, the number of levels constituting the surface transition has been considered

to be fixed and small. Later considerations will show that we would expect the interface to be correctly represented by certainly more than two layers. The models of two- and three-layer surfaces have been adopted for simplicity but it was noted that Burton *et al.* showed that going to more than three layers in their model did not radically change the form of the behaviour of the transition. An alternative approach to the problem is that of Temkin (1964) who considers the number of layers involved in the transition to be one of the variables which can adjust itself to minimise free energy. This is, of course, a far more realistic model, although even a model of a fixed number of layers can equilibrate in a configuration having some of the layers all liquid or all solid; that is, a model of a fixed number of layers can *reduce* that number. One of the problems which might be expected to be associated with this broader view is that the problem will not be solvable exactly. However, this is true of any model involving more than two layers. The approximations used by Temkin to solve this problem are the Bragg–Williams ones used already and we shall now investigate this theory in rather more detail.

Temkin's model is shown schematically in fig. 3.5. For simplicity the calculation is restricted to the (100) surface of a simple cubic material.

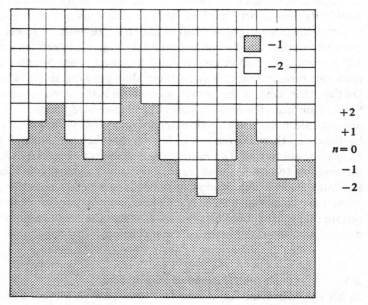

Fig. 3.5. Schematic diagram of Temkin's model of the interface.

Liquid atoms are effectively assumed to be in the positions similar to those of solid atoms on a lattice but are in the liquid state (see fig. 3.5). Clearly this is a gross simplification in that it is obviously not a real representation of the instantaneous state of the system, but this is probably not too serious over a time average which is all we are concerned with here. Also, the model assumes that any 'solid atom' has another solid atom immediately below it in the solid; that is, in the phraseology of §3.1, it does not allow for the possibility of 'overhangs' in the 'mountain range structure'. The layers comprising the transition are labelled by a number n and the $n = 0$ layer is approximately at the mean position of the interface (though this mean position is not supposed, *a priori*, to be *exactly* at the $n = 0$ layer). This is ensured by imposing the boundary condition that if there are N_n 'solid atoms' in the nth layer which can accommodate N atoms, then writing $C_n = (N_n/N)$, we require that $C_{-\infty} = 1$ and that $C_{+\infty} = 0$.

The procedure is now simply to determine the total free energy of the system $F(N, C_n, T)$ depending on all the C_n and temperature. Notice that the proper statistical mechanical treatment generates the Helmholtz free energy, F, rather than the Gibbs free energy, G. However, for the process of melting which occurs approximately at constant volume, the two are equivalent. This is done in the rather more general way of writing

$$F(N, C_n, T) = -kT \ln Z(N, C_n, T) \tag{3.14}$$

where Z is the partition function. This can be evaluated if we know the internal energy of all possible states of the system. To do this we split the internal energy of each atom into the sum of the bond energies between two 'solid atoms', two 'liquid atoms' and 'solid' and 'liquid' atom pairs ϵ_{11}, ϵ_{22} and ϵ_{12}. Then the contributions to the energy of the nth layer due to its interaction with the $(n - 1)$th and $(n + 1)$th layers, respectively, are

$$\left. \begin{aligned} E_{n,n-1} &= (N/2)[C_n\epsilon_{11} + (1 - C_{n-1})\epsilon_{22} + (C_{n-1} - C_n)\epsilon_{12}], \\ E_{n,n+1} &= (N/2)[C_{n+1}\epsilon_{11} + (1 - C_n)\epsilon_{22} + (C_n - C_{n+1})\epsilon_{12}]. \end{aligned} \right\} \tag{3.15}$$

(Note the assumption made here that each solid atom has another nearest neighbour solid atom on its bulk solid side.) The energy of the nth layer due to interactions of atoms within the layer is

$$E_{n,n} = N_{11,n}\epsilon_{11} + N_{22,n}\epsilon_{22} + N_{12\ n}\epsilon_{12}, \tag{3.16}$$

$N_{11,n}$ being the number of solid–solid bonds in the nth layer, etc. As there are a total of $2N$ bonds within each layer, these Ns must satisfy

and
$$\left.\begin{array}{l} N_{11,n} = 2NC_n - (N_{12,n}/2) \\ N_{22,n} = 2N(1 - C_n) - (N_{12,n}/2). \end{array}\right\} \qquad (3.17)$$

Thus (3.16) can be written as

$$E_{n,n} = 2NC_n\epsilon_{11} + 2N(1 - C_n)\epsilon_{22} + N_{12,n}w \qquad (3.18)$$

where

$$w = \epsilon_{12} - \frac{(\epsilon_{11} + \epsilon_{22})}{2}. \qquad (3.19)$$

In order to evaluate the partition function it is now necessary to make some simplifications to eliminate the $N_{12,n}$ terms, or at least to make this a well-defined unique function of n. The Bragg–Williams approximation is used which is to assign to each N and C_n some mean value of $\bar{N}_{12,n}$. The value chosen is the simple unweighted mean

$$\bar{N}_{12,n} = 4NC_n(1 - C_n). \qquad (3.20)$$

(This is the same approximation used by Jackson in writing down (3.5). In addition Jackson effectively assumes that $\epsilon_{22} = \epsilon_{12}$.) By writing down the partition function, Z_0, for a smooth boundary (i.e. $C_n = 1$ for $0 \geqslant C_n \geqslant -\infty$, and $C_n = 0$ for $\infty \geqslant C_n \geqslant 1$), it is possible to eliminate many terms (which for an infinite solid would be infinite) by evaluating the excess free energy of the random configuration over that for the smooth interface as

$$\Delta G = -kT \ln (Z/Z_0). \qquad (3.21)$$

The result obtained is

$$\frac{\Delta G}{NkT} = \beta\left[\sum_{n=-\infty}^{0}(1 - C_n) - \sum_{1}^{\infty}C_n\right] + \gamma\sum_{n=-\infty}^{\infty}C_n(1 - C_n)$$
$$+ \sum_{n=-\infty}^{\infty}(C_{n-1} - C_n)\ln(C_{n-1} - C_n) \qquad (3.22)$$

where $\beta = (\Delta\mu/kT)$, $\gamma = (4w/kT)$ and $\Delta\mu$ is the difference in chemical potential between atoms in the two phases. Thus, at equilibrium $\beta = 0$, and is otherwise proportional to ΔT the supercooling (or superheating). We shall restrict ourselves to the equilibrium ($\beta = 0$) solutions in this chapter; non-equilibrium situations will be dealt with in §8.4. Equation (3.22) may be compared with Jackson's final result in (3.13). Using the same approximations as Jackson ($\epsilon_{22} = \epsilon_{12}$) and applying Jackson's crystallographic factor (η_1/ν) to the (100) surface of a simple cubic material shows that $\alpha = \gamma = \frac{2}{3}(L_0/kT_E)$. Thus α and γ are essentially the same quantity. Now reducing the summations of (3.22) to the single disordered layer in Jackson's model, we see that this equation is identical to (3.13).

However, it is clearly not possible to plot (3.22) in the simple way used for (3.13); there are still an infinite number of variables, the C_ns. Differentiating (3.22) and equating to zero to find the equilibrium (minimum free energy) configuration gives

$$\frac{C_n - C_{n+1}}{C_{n-1} - C_n} \exp (2\gamma C_n) = \exp (-\gamma + \beta). \tag{3.23}$$

Temkin solves these equations (3.23) numerically and for equilibrium arrives at two possible types of solution

(I) $C_0 = 1 - C_1$, $C_{-1} = 1 - C_2$,   etc.   (3.24)

(II) $C_0 = \frac{1}{2}$, $C_{-1} = 1 - C_{+1}$,   etc.   (3.25)

A check on the second derivative of (3.22) for these solutions shows that only solution (I) is in fact a true minimum in the $G(C_n)$ function and that (II) corresponds to a saddle point in this function and is therefore presumably not a stable configuration. Thus the analysis concludes that for all values of γ (or α) there is a minimum free energy configuration when the mean interface position lies midway between the $n = 0$ and $n = 1$ layers. The excess free energies of the arrangements given by the two solutions (3.24) and (3.25) are shown in fig. 3.6 as a function of $1/\gamma$. Further, Temkin has computed the values of C_n in the interface for various values of γ and these are shown in fig. 3.7. Notice that a typical value for a metal of $\gamma = 1$ (for a low index surface) will have an interface spread over about 6 or 8 atomic layers, whereas the $\gamma = 3.3$ plot which would correspond to a markedly faceting material represents an interface thickness of only 2 or 3 atomic layers.

Finally, the form of the roughness dependence on γ can be appreciated by looking at the curve of surface roughness (using the definition of Burton *et al.*) against $1/\gamma$. For this model, the surface roughness is given by

$$s \equiv \frac{\Delta E_s}{Nw} = 4 \sum_{n=-\infty}^{+\infty} C_n(1 - C_n) \tag{3.26}$$

(the parameter s for the Jackson model is simply the one term of this sum resulting from the one disordered layer model). The relationship is shown in fig. 3.8 and may be compared with the results of the other models in fig. 3.4; by extending the model beyond two levels the sharp

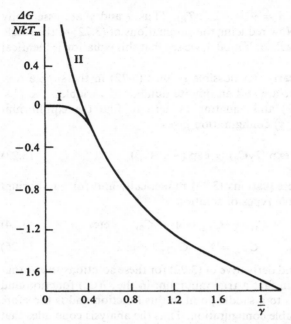

Fig. 3.6. Excess free energy versus $1/\gamma$ for solutions of type I (equation (3.24)) and II (equation (3.25)).

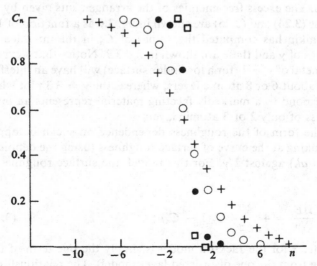

Fig. 3.7. Concentration distribution of 'solid' atoms C_n at equilibrium for various values of γ.
$+\ \gamma = 0.446$, $\bigcirc\ \gamma = 0.769$, $\bullet\ \gamma = 1.889$, $\square\ \gamma = 3.310$.

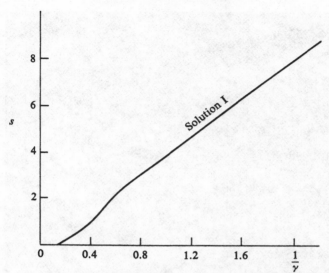

Fig. 3.8. Surface roughness s versus $1/\gamma$ for Temkin solution.

singularity has been lost as expected and now there is simply a point of inflection at $1/\gamma \simeq 0.41$ compared with the singularity in the Jackson model at $1/\alpha = 0.5$.

The general conclusions of this multilayer model are therefore for the equilibrium ($\beta = 0$) situation, the conclusions arrived at earlier in the chapter. Thus materials having $\alpha > 2$ generally have much smoother interfaces than those with $\alpha < 2$, though the transition between the two is far less well defined than the two-level model would suggest. The non-equilibrium ($\beta \neq 0$) situation will be dealt with in chapter 8. One interesting feature which emerges from this analysis is that the two-level model used by Jackson means that for rough materials the minimum free energy configuration given by fig. 3.1 is in fact a solution of the metastable type (equation (3.25)) and is therefore not a true equilibrium configuration; this is obviously an artefact of the over-restricted model and its implied boundary conditions.

3.6   Interface morphologies during melting

While the discussion of the preceding sections has related to the faceted and non-faceted morphologies of the interface during solidification, the theory has been essentially based on equilibrium situations. Thus it might be reasonable to suppose that the property of a material to facet or not would be displayed both in solidification and melting. Certainly,

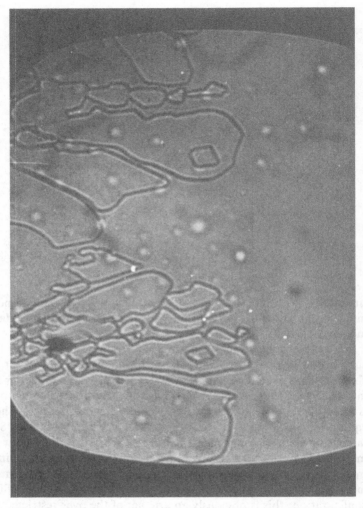

Fig. 3.9. Salol melting in a temperature gradient (increasing from left to right), × 235. Melting within a single crystal is seen to show a faceted interface. (After Woodruff, 1968b.)

if the *principle of microscopic reversibility* implied by many early crystal growth workers (Tammann, 1925) that atom movements at the interface during solidification are essentially reversed during melting is accepted, then combined with the equilibrium theories this conclusion seems inevitable. Thus it is perhaps surprising to find that if, for example, the temperature of a melt in which faceting crystals are growing is increased to above the melting point, all the corners of the crystals melt off and they melt away with a rounded interface typical of low α-factor materials in growth. Similarly, in a uniaxial growth configuration the corners of a faceted solidifying interface melt away when the temperature is raised, and melting continues on a rounded interface with liquid channels penetrating between crystals along grain boundaries.

However, if localised melting occurs *within* a single crystal of a faceting material, the melting interface is seen to facet parallel to those faces formed on the same crystal during solidification (e.g. fig. 3.9). If these faceted pools are refrozen, solidification rounds off the interface at the corners of the pool and freezing continues on a rounded interface. Other observations show that if internal melting occurs in a non-faceting material, the observed melting pools are also non-faceted (Woodruff, 1968*b*; Woodruff & Forty, 1967).

In fact these observations can be readily understood if, in the case of faceting materials, growth of *either phase* will occur easily on an atomically rough interface only while the geometrical conditions imposed on the system prevent their formation. Otherwise facets will form and the phase will then continue to grow with a faceted interface. This is certainly true in the case of solidification. If a faceting crystal is partially melted so that it displays a rounded interface, on resolidifying growth rapidly occurs on the rounded high index orientation parts (which are presumably atomically rough) until no further growth can occur without it taking place on the low index orientation surfaces; i.e. until facets have formed. On the other hand, if uniaxial solidification of a faceting material is carried out under a temperature distribution such that the interface is concave relative to the solid (and falls behind the line the interface would take if facets were to form) then growth occurs on a non-faceted interface (Hulme & Mullin, 1962). In this case the imposed conditions prevent growth from moving the interface out to the facet plane positions. For these reasons also, melting of a crystal within a melt will occur in a non-faceted manner because the interface is concave relative to the growing phase which in this case is the melt. However, internal pools of liquid have a convex interface relative to the melt as they grow and so will be expected to show a faceted interface if the material facets during solidification.

These comments show, therefore, that there is in fact a symmetry in the cases of melting and solidification, and that the 'principle of microscopic reversibility' does seem to be essentially valid.

3.7   The structure of the interface in alloy systems

While all the models discussed so far are restricted to pure materials, systems of practical interest are often alloys and it is therefore pertinent to ask whether any special considerations should be given to these cases. To this end Taylor, Fidler & Smith (1968) have extended Jackson's model to certain binary alloys. In view of the apparent success of Jackson's theory to pure materials, and its relative simplicity, this is a reasonable starting point for a first attempt at introducing extra complicating factors associated with alloys. It may, of course, be tempting providence to take a model which is admittedly naive and expect it to stand up to extension to even more complicated systems, but no doubt this should be manifested in the results if this turns out to be the case.

The analysis follows Jackson's model exactly and the basic equation for the excess free energy of the partially filled monolayer (cf. (3.1)) in the alloy of A and B atoms is

$$\Delta G = -\Delta E_{0A} - \Delta E_{0B} - \Delta E_{1A} - \Delta E_{1B} + T\Delta S_\alpha$$

$$- T\Delta S_1 + P\Delta V \tag{3.27}$$

where $-\Delta E_{0A}$ and $-\Delta E_{0B}$ are the energies gained by adding A and B atoms respectively from the liquid solution due to the interaction with atoms within the solid below and $-\Delta E_{1A}$ and $-\Delta E_{1B}$ are the similar terms relating to energies of interaction of A and B atoms with other atoms in the added layer. ΔS_α is the entropy of fusion of the α-phase of the solid (i.e. the solid freezing out of the melt); this may be alternatively thought of as an entropy of solution as the adjacent equilibrium melt will of course have a different composition in general. ΔS_1 is simply the configurational entropy associated with the different possible sitings of the $N_A + N_B$ atoms in the added layer (but does not include the configurational entropy of mixing the N_A and N_B atoms; this is included automatically in ΔS_α). ΔV is the volume change as before. The method of calculating these terms from fairly well-known parameters has been given by Kerr & Winegard (1967) with some further corrections by Taylor *et al.* The final solution turns out to be as for Jackson's analysis except that L_α replaces L_0 in the new α-factor being the heat of solution or fusion of the α-phase.

Taylor *et al.* then apply their result to the silver–bismuth alloy system. The results are shown for the silver-rich phase in fig. 3.10, in the form of excess free energy versus occupation of the partially filled layer, for various different alloy compositions at their respective equilibrium temperatures. The interesting point is that while silver has an α-factor well below 2 in its pure state, these curves show that over a certain range of melt compositions the silver-rich phase has an α-factor in excess of 2. In particular, this phase should be faceting near the eutectic composition (see chapter 7), but should become non-faceting as the melt composition

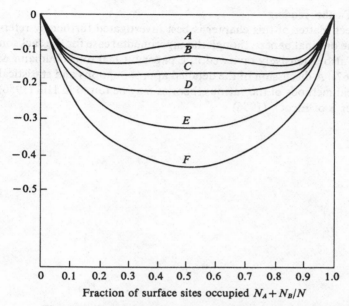

Fig. 3.10. Relative surface free energy as a function of monolayer occupation for (111) surface in the silver-rich phase of silver–bismuth (after Taylor, Fidler & Smith, 1968).

Curve	Liquidus temp. (0° C)	Composition of melt (at. %Ag)	Roughness parameter α
A	251	3.2 (undercooled)	2.39
B	262	4.7 (eutectic)	2.27
C	280	7.1	2.09
D	300	9.7	1.90
E	350	18.1	1.52
F	400	37.5	1.02

approaches about 9 atomic per cent silver. While the exact position of the transition is not clear, Taylor *et al.* show that such a transition does indeed occur and that rounded dendritic structures are observed at high silver content and faceted dendrites appear near the eutectic composition. The success of this result may be offset to some extent by remarking that this type of analysis would also predict non-faceted growth of the bismuth-rich phase and of pure bismuth ($\alpha \simeq 1.6$) whereas pure bismuth and the bismuth-rich phase of alloys, of low silver composition at least (Miller & Chadwick, 1968) seem to grow with a faceted interface. Of course, this is a more borderline case than the transition in silver so that the prediction may still be considered a notable success.

Further reading

The subject matter of this chapter is best investigated further by reference to the original papers, though the experimental case for the Jackson theory is also extensively reviewed in a paper by Jackson, Uhlmann & Hunt (1967). A discussion of the detailed approximations and statistical mechanical methods of the various theories may be found in Hill (1956) or chapter 3 of Elcock (1956).

4   Non-structural views of the solid–liquid interface

4.1   Introduction

The purpose of this chapter is to bring together the various theoretical studies which cannot be classified as structural theories but which are nevertheless intended to improve our understanding of the solid–liquid interface. These consist of certain theories of melting itself, theoretical developments of the idea of a grain boundary as a liquid region which provide one method of estimating the thickness of the interface, and a small amount of work aimed at calculating the free energy of this interface.

Firstly, however, we investigate briefly an alternative model of the interface which considers the 'diffuseness' rather than the 'roughness' of the interface. It therefore avoids the idea utilised in the structural theories of chapter 3 that individual atoms within the interface region can be ascribed to one phase or the other. While the interface is still supposed to have some well-defined thickness it is no longer possible to say that a particular atom within this region belongs only to the liquid or the solid in a particular position or structure, but rather that the atomic distribution becomes more liquid-like or solid-like as one passes through the boundary. This is probably a more realistic picture of the interface in view of the constant motion and interchange of atoms between solid and liquid phases at the interface, although it may not serve as convenient a basis in the construction of a proper theory of crystal growth.

The idea of describing the interface as diffuse rather than rough has been examined theoretically in another treatment by Temkin (1966). This is done formally by using an order–disorder theory approach in which it is supposed that atoms can lie on one of two sub-lattices α or β. For example, where the solid is face-centred cubic, one sub-lattice is given by the normal solid structure and the second by the centres of the cube edges. Now at absolute zero in temperature when the solid is completely ordered all atoms lie on the points of one of these sub-lattices – say the α one. As the temperature is raised a certain amount of disordering is introduced by allowing some atoms to move to points on the β sub-lattice. In the completely disordered state (in our case the liquid) equal numbers of α and β sites are occupied. Now the model can be used to describe the solid–liquid interface by allowing some disorder parameter to change from the solid to liquid values over a certain number of lattice spacings; if this number can be deduced from free energy minimisation then the 'interface thickness' is deduced. This is done formally by introducing a long-range order parameter

$$y = \frac{2N_\alpha}{N} - 1 \qquad (4.1)$$

where N_α is the number of atoms at α sites and N the total number of atoms in the layer. Thus for complete order $y = 1$ (or -1 if all atoms are at β sites) and for total disorder $y = 0$. Temkin (1966) does not carry out the full analysis and comparison for the solid–liquid interface, but concludes that the results are generally in agreement with his rough interface conclusions (Temkin, 1964 – see §3.5). Thus he concludes that the thickness of the boundary is an inverse function of the entropy of fusion, and in fact also finds the two possible equilibrium solutions (stable and metastable) for the mean position of the interface relative to the atom planes. The general conclusions on the non-equilibrium situation are also in agreement with his rough interface analysis (see chapter 8).

4.2   Theories of melting

Most of the theories of melting which have been proposed over a long period of time are essentially phenomenological and are based on the observation that certain parameters of the solid seem to reach a critical value at the melting point. For example, a common idea is that as a solid is heated the atoms start to vibrate with greater and greater amplitudes until, at some critical value (around 7–10 per cent of the lattice spacing), the interaction of adjacent atoms 'shakes the solid to pieces'. This, in rather crude terms, is the basis of the Lindemann (1910) theory and also of many other more refined versions of the same idea. Other effects of this kind which have been remarked upon are critical concentrations of vacancies reached at the melting point and the disappearance of certain mechanical strength parameters. Unfortunately, all these theories suffer from certain common inherent difficulties. Firstly, they fail to answer properly the question of why these particular conditions should be so critical, and secondly, they imply that at the appointed temperature the whole crystal should collapse. In other words, any superheating of solids should be impossible and the transformation should be spontaneous rather than a process involving nucleation and growth. These theories are therefore of little interest in studying the solid–liquid interface in that they largely deny the necessity even for the existence of such an interface.

A more relevant attempt to provide a theory of melting has been made by Kuhlmann-Wilsdorf (1965) on the basis of crystal dislocations. This also allows one to calculate the solid–liquid interfacial free energy. The role of dislocations in melting had been considered previously by several other authors but this latest work is most suitable for our present discussion. The idea is basically as follows. The free energy of a dislocation is the sum of a long-range elastic distortion energy plus a contribution due to the core.† This is the central part of the dislocation line a few lattice spacings in diameter for which the distortion is too great to be treated by an elastic continuum theory; indeed such a theory would imply infinite distortion and so infinite energy at the centre of the dislocation line. The total energy is of the order of 10 eV per lattice plane threaded by the dislocation which is much higher than could be provided by thermal excitation. It is clear therefore that dislocations are not normally in thermal equilibrium with the crystal lattice. However, the main contribution to the energy arises from the long-range elastic distortion and if two dislocations of opposite Burgers vector are brought close together most of this distortion is eliminated. Thus narrow dislocation loops *can* be in thermal equilibrium with a lattice because their energy is almost entirely due to the much smaller contribution from the dislocation cores alone. Kuhlmann-Wilsdorf has calculated the vibrational entropy and hence the total free energy of dislocation cores; the internal energy has been estimated by many authors and Cottrell (1953) has also shown that the configurational entropy is much less than the vibrational entropy. Kuhlmann-Wilsdorf has shown that the free energy will vanish at some well-defined temperature when the entropy term becomes equal and opposite to the internal energy term. At such a temperature, dislocation cores should be generated spontaneously throughout the crystal; when this happens the crystal will lose all its shear strength and the transition will be accompanied by an absorption of latent heat. The temperature at which this occurs turns out to be close to the experimentally observed melting point and it would appear therefore that the transition predicted is that of melting. In fact, in determining the melting temperature, latent heat and entropy of melting from the theory, certain parameters are needed which are not well known but which can be deduced from observed values and then shown to be mutually consistent. Agreement with experiment is quite good in

† For an introduction to dislocation theory see a standard text on this subject such as Cottrell (1953) or Read (1953).

this respect. Moreover, the theory can be shown to agree well with some of the empirical relationships for the melting transition mentioned earlier; for example, Lindemann's (1910) theory relates the melting point to the Debye temperature of the solid, which, because of the relationship between this and the elastic constants of the solid can be shown to be compatible with the dislocation theory.

A possible source of error in this approach is the assumption that small dislocation loops have no long-range elastic energy. In fact this is not completely true for an isolated loop or dipole and, as the elastic component of the energy is very much greater than that of the core, it is clear that a small error in this approximation would have a large effect on the plausibility of the foregoing argument. However, it should be noted that for the new state to be really liquid-like large groups of closely spaced dipoles must be generated, in which case the elastic energy within these groups should be truly zero. Outside these groups elastic strain energy still exists; in other words there is an interfacial free energy between the two phases, and an initial nucleation problem exists in that melting must begin with the nucleation of groups of closely spaced dislocation core dipoles of a critical size. Thus the theory has generated a true nucleation and growth theory.

Kotzé & Kuhlmann-Wilsdorf (1966) have shown that on the basis of these ideas the value of the solid–liquid interfacial free energy can be calculated from the energy associated with the edge of these dislocation arrays. In fact they deduce that the free energy should be half the free energy of a general dislocation grain boundary, which can be calculated from the elastic constants of the material. Agreement with experimental results from homogeneous nucleation experiments is quite striking with the ratio of theoretical and experimental values varying between about 0.80 and 1.25 (see fig. 4.1).

Finally, the theory makes one interesting prediction. Kuhlmann-Wilsdorf points out that the dislocation core dipoles are probably generated from glide motions of dislocation, and in particular, from dislocation nodes, which are likely to be present in the crystal already. This would therefore imply that, for most materials in which such dislocations are almost invariably present, significant superheatings are unlikely to be observable due to the ease of generation of the melting nuclei. For dislocation-free materials this may not be true. Certainly those materials in which superheating has been observed seem to be ones in which the dislocation properties are not well understood and are likely to be complex. It is perhaps notable also that gallium which appears to be a material which can be easily superheated by small amounts is very easy to grow in dislocation-free form (Pennington, 1966).

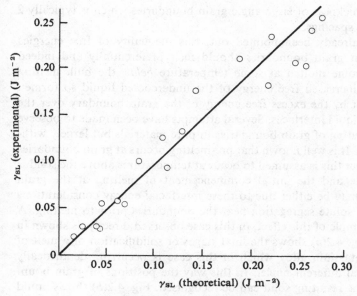

Fig. 4.1. Comparison of experimental values of γ_{SL} from Turnbull's nucleation experiments and the theoretical values from Kotzé & Kuhlmann-Wilsdorf (1966).

4.3   The thickness of the solid–liquid interface

One possible way of arriving at an empirical value for the thickness of the solid–liquid interface comes from observations on grain boundaries. It has been a popular view for many years that one way of considering a grain boundary as a disordered region between two crystals is to think of it as a thin layer of liquid; obviously this disordered state is liquid-like though whether it is useful to use ideas of bulk liquids to describe it when the grain boundary thickness is only a few atomic spacings is subject to argument. However, assuming that this is a useful viewpoint, it follows that the free energy of a grain boundary should simply be twice that for the solid–liquid interface. Indeed this is the conclusion arrived at by Kotzé and Kuhlmann-Wilsdorf by quite different reasoning. However, we have already seen in chapter 2 that experiments on pure materials seem to show the inequality $\gamma_{GB} > 2\gamma_{SL}$. Now, on the basis of the liquid model of grain boundaries this would be impossible if the solid–liquid interface were sharply defined. The extra free energy of the grain boundary must be due to the fact that the two solid–liquid interfaces interact; i.e. that the two interfaces are starting to overlap. Thus we deduce that the interface thickness is of the order of the observed thickness of large angle grain boundaries which is typically 2 or 3 lattice spacings.

As has already been pointed out, this inequality of free energies implies that grain boundaries should melt preferentially and indeed should become molten at some temperature *below* the bulk melting point, the increased free energy of the undercooled liquid so formed being offset by the excess free energy of the grain boundary over the two solid–liquid interfaces. Several attempts have been made to observe this pre-melting of grain boundaries in pure materials but largely without success. It is well known that pre-melting occurs at grain boundaries in alloys, but this is assumed to occur at temperatures above the solidus temperature and the initial commencement of melting at the grain boundary may be either due to these interfacial energy considerations or to local solute segregation near the boundaries prior to melting. A typical example of this effect, in this case observed directly, is shown in fig. 4.2. Fig. 4.2(*a*) shows the final stages of solidification of a melt of sodium–potassium alloy which in this case is occurring dendritically from several different nuclei. In this way the positions of grain boundaries in the resulting solid can be recognised. Fig. 4.2(*b*) shows liquid regions forming preferentially at these grain boundaries when the specimen is subsequently re-melted. These observations were made using transmission ultra-violet microscopy (Forty & Woodruff, 1969). Similar effects, involving segregation of unknown impurities probably account for most of the observations of pre-melting in so-called pure materials.

A proper theoretical treatment as to whether grain boundary melting should be observable is not really possible. If the solid–liquid interface were sharply defined, then an upper limit for the extent of melting could be established by balancing the increased bulk-free energy of the super-cooled liquid against the drop in interfacial free energy. However, if the solid–liquid interface were absolutely smooth a lower free energy state would be achieved if melting did not occur and indeed the minimum free energy configuration would be a single atomic layer of liquid between the two interfaces maintained until the melting point was reached. The finite thickness of the interface means that the grain boundary energy must be composed of the sum of the solid–liquid interfacial free energies plus some term due to the interaction of the two interfaces. This may be written by saying that the grain boundary energy must be composed of contributions due to the two solid–liquid interfaces, the excess free energy of the undercooled 'liquid' and an extra part due to the overlapping of the interfaces. Thus

$$\gamma_{GB} = 2\gamma_{SL} + nL_0 \frac{\Delta T}{T_m} V^{-2/3} + \Phi(n)V^{-2/3} \tag{4.2}$$

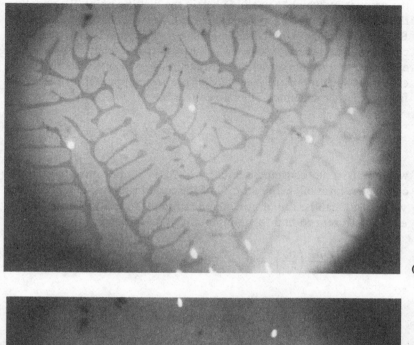

(a)

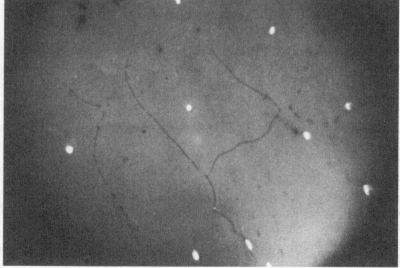

(b)

Fig. 4.2. (a) Final stages of dendritic freezing from various nuclei, and (b) subsequent melting of grain boundaries in the same specimen of potassium–sodium alloy. Photographs taken by transmission ultra-violet microscopy × 95.

where n is the number of atomic layers comprising the boundary, ΔT is the amount by which the solid is below the melting point T_{m}, and V is the atomic volume (L_0 being the latent heat per atom). Following Bolling (1968) and Uhlmann, Chalmers & Jackson (1964), who were considering a related problem, we may suppose that $\Phi(n)$ can be expressed as a simple power law proportional to $(n_0/n)^p$ where n_0 is the constant which represents the 'closest probable approach of the interfaces' in lattice spacings. Thus we can write

$$\Phi(n) = \left(\frac{n_0}{n}\right)^p L_0 \phi \tag{4.3}$$

where ϕ is some constant number and L_0 is simply introduced as a convenient form of energy unit. Now the equilibrium value of n for any value of ΔT can be found by differentiating (4.2) with respect to n and equating to zero to find the minimum of the expression for γ. Thus

$$\frac{\Delta T}{T_{\mathrm{m}}} = \frac{p\phi}{n_0}\left(\frac{n_0}{n_{\mathrm{eq}}}\right)^{p+1}, \tag{4.4}$$

so

$$\frac{n_{\mathrm{eq}}}{n_0} = \left(\frac{p\phi}{n_0}\frac{T_{\mathrm{m}}}{\Delta T}\right)^{1/p+1}. \tag{4.5}$$

Bolling has shown that the choice of $p = 3$ and $\phi = 2.6$ gives results which are consistent with our knowledge of the energy and dimensions of grain boundaries. n_0 is assumed to be unity (i.e. the interfaces can never approach closer than the normal atomic spacing). A general value for γ_{SL} can be taken from Turnbull's empirical law given in chapter 2 leading to

$$\gamma_{\mathrm{SL}} = 0.46 L_0 V^{-2/3}. \tag{4.6}$$

Fig. 4.3 shows the equation (4.2) with $\Phi(n)$ expressed in terms of (4.3) plotted for three values of $\Delta T/T_{\mathrm{m}}$. Notice the very shallow minima which occur at higher temperatures. In particular we see that the values of n_{eq} found for temperatures in the range $0.5\,T_{\mathrm{m}}$ to $0.9\,T_{\mathrm{m}}$ are between two and three in good agreement with general ideas about the structure of grain boundaries and evidence from field ion microscopy (see, for example, Muller & Tsong, 1969). Also the total grain boundary energy γ_{GB} tends to $L_0 V^{-2/3}$ as $\Delta T \to 0$. From (4.6) this means that $\gamma_{\mathrm{GB}} \to \gamma_{\mathrm{SL}}/0.46$, which is in excellent agreement with the experimental evidence of Miller & Chadwick (1967) that $\gamma_{\mathrm{GB}} \to \gamma_{\mathrm{SL}}/0.45$ at the melting point.

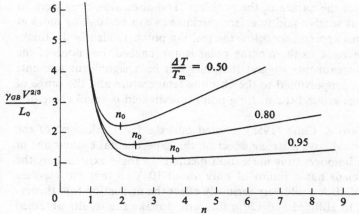

Fig. 4.3. Schematic graph of the total interfacial free energy of a system of two solid–liquid interfaces separated by n lattice spacings (after Bolling, 1968).

Using these constants and (4.5) some values of n_{eq} can be evaluated for other values of $\Delta T/T_m$:

$$\Delta T/T_m = 0.5, \qquad n_{eq} = 1.99;$$

$$\Delta T/T_m = 0.05, \qquad n_{eq} = 3.54;$$

$$\Delta T/T_m = 10^{-5}, \qquad n_{eq} = 29.8.$$

The calculation is fairly insensitive to the value of p chosen.

These results imply that for a material of melting point $1000°K$, the molten layer thickness would be only about 100 Å at only $1/100$ deg below the equilibrium melting point. An increase of n_0 to 2 which is probably as large as is feasible to still give reasonable values of grain boundary thicknesses at low temperatures, would only double these values of n_{eq}.

On this basis the molten thickness would not reach the thickness of one micron until the temperature was 10^{-13} deg from the melting point. It is therefore not surprising that grain boundary melting has not been observed. Some attempts have been made to detect an effect indirectly by looking for an associated loss of mechanical strength of the material, but as Miller & Chadwick (1967) pointed out, a very thin liquid layer can provide excellent adhesion between the grains. Indeed, these workers estimated that if this thin layer of liquid behaved like bulk liquid, then such a loss of strength would not be detected until the melted layer was of the order of 10 microns thick. Of course all the foregoing analysis is very crude and rather speculative, but it certainly demonstrates the nature of the problem. The discussion does seem to indicate that solid–liquid interface thicknesses can be 10 Å or more at temperatures appreciably below the melting point. Evidently the thickness will increase as the melting point is approached, but none of the theoretical treatments suggest that this will be a significant amount; the effect is proportional to the absolute temperature and the range of temperatures around the melting point investigated is small compared with this.

As Hilliard & Cahn (1958) pointed out, the finite thickness of the interface could have a strong effect on the experimental evaluations of γ_{SL} by the homogeneous nucleation method. In these experiments the critical nucleus has a radius of only about 10 Å so that an interface thickness of 10 Å could very seriously affect the nucleation rate theory. According to Hilliard and Cahn this would make the results obtained by this method too small. It is therefore surprising that the correlation with other experiments is so good. Admittedly, many results obtained in this way can be related to grain boundary energies in the same material by a rule $\gamma_{SL} \simeq 0.30\gamma_{GB}$ which is much lower than Miller and Chadwick's direct measurements, but this rule is derived from measurement of γ_{GB} well below the melting point while Miller and Chadwick's are extrapolated to T_m. One obvious possibility is that as nucleation experiments are performed at temperatures well below the melting point, the temperature dependence of the free energy via the interfacial entropy may offset this effect. No experimental values are available but theoretical arguments suggest that the entropy term may be small (Skapski, 1956; Kotzé & Kuhlmann-Wilsdorf, 1966).

Thus, in the absence of more extensive experimental data from other techniques the extent of the discrepancies cannot be resolved and so no further comment is worthwhile. It is notable, though, that the theoretical treatment of Kotzé and Kuhlmann-Wilsdorf, based on a grain boundary relation but calculated from the elastic constants, gave much better agreement with the nucleation experiment results (fig. 4.1) than Miller and Chadwick's comments would suggest.

A possibility that is related to the problem of grain boundary melting as a result of the decrease in interfacial free energies is the occurrence of melting at a free surface at a temperature below the bulk melting point. This was mentioned earlier in chapter 2. In the case of most solid surfaces $\gamma_S > \gamma_{SL} + \gamma_L$ so that there should be a driving force for a thin surface molten region to form at the surface below the melting point. On the other hand it is well known from low energy electron diffraction studies that close-packed surfaces of a solid are certainly well ordered and no disordered layer even one monolayer thick exists on the surface. Thus, unlike the grain boundary problem there is no obvious choice of n_0 available; indeed, in order for a molten zone of a finite thickness to form it should first be necessary for the surface mono-layer to become disordered. This is, of course, the so-called 'surface melting' effect of Burton, Cabrera and Frank. However, it is not clear how far this kind of argument is valid; Burton *et al.* suggest that 'surface melting' will not occur, below the melting point, on the close-packed planes of cubic materials but it is fairly certain that the fore-mentioned inequality of interfacial energies applies to such a surface. At the same time, the analysis applied to grain boundary melting above shows that even if two-dimensional 'surface melting' allowed a simple choice of n_0, no gross three-dimensional surface melting should be observable at temperatures below the melting point by more than a small fraction of a degree.

4.4   Theoretical calculations of $\gamma_{\rm SL}$

The normal method of calculating interfacial energies for solid surfaces involves summing the energies associated with 'broken' or 'dangling' bonds due to the absence of similar atoms on the liquid or vapour side of the interface. In principle all 'bonds' between atoms should be included but it is common to make the assumption that as the inter-action forces are short range only the bonds between nearest neighbour atoms need be considered. There are then two alternative approaches to the solution; either to simply associate with the broken bonds a suitable proportion of the binding energy of each atom in the solid (rather like the models for rough interfaces considered in the last chapter), or to consider specific interatomic potentials with the help of one or more adjustable parameters. Common potentials to choose are those due to Morse or Mie and there are several examples of such calculations for the solid–vapour or free solid interface (e.g. Nicholas, 1968). This approach has the advantage of being more useful in evalua-ting the variation of γ with orientation as extra neighbour interactions can easily be included. Such an approach does not seem to have been developed for the solid–liquid interface, but Skapski (1956) has at-tempted a calculation based on the alternative nearest neighbour bonding approach. In his calculation he uses the device of overcoming possible systematic errors by calculating the solid–liquid interfacial free energy as a function of the surface tension of the free liquid surface, which is rather well known experimentally. In fact he really derives the free solid surface energy and the value for the solid–liquid interface is arrived at via the equation

$$\gamma_{SL} = \gamma_S - \gamma_L. \tag{4.7}$$

This is not strictly true in general. Indeed it has already been pointed out in §2.5 (see (2.29)) that for most materials and surfaces

$$\gamma_{SL} < \gamma_S - \gamma_L. \tag{4.8}$$

For certain other cases it is possible that the inequality is reversed. However, (4.7) is probably a useful approximation, and in any case the correlation of the values for γ_{SL} calculated in this way with experimental results from undercooling experiments is quite reasonable (see table 4.1). In this case the theoretical values are for the closest packed surfaces where the free energy is lowest.

Table 4.1

	Surface tension γ_{SL} (10^{-3} J m^{-2})		
	Experimental		Theory
Metal	(Turnbull)	(Skapski)	(Zadumkin)
Na	20†	15	11.2
Li	30†	27	—
Pb	33	32	25.6
Ag	126	101	85
Au	132	121	110
Cu	177	128	132
Pt	240	236	230

† J. W. Taylor, *Phil. Mag.* **46**, 857 (1955).

A rather different approach to the problem has been given by Zadumkin (1962) who calculates the values for metals by considering the electron–ion core interaction. This should of course be equivalent to a bonding calculation taking into account all neighbour interactions but using the specific form of interaction in the metallic state. He concludes, after summing the various contributory terms to the energy, that the main contribution is simply due to the volume change (and hence density change) on melting. Various significant approximations are included in the theory; for example, to aid the calculation it is assumed that the change in electron density on melting is due entirely to the change in mass density and that there is no change in the average number of 'free electrons' per ion core (an assumption not supported by experimental results on electronic properties of metals).

This analysis is related to an earlier one by the same author for the

free solid surface and he also relates these results to the liquid state
surface tension, in this case using the assumption that

$$\{\gamma_S(hkl)\}_{min} \simeq \gamma_L, \tag{4.9}$$

i.e. that the liquid surface tension is equal to the lowest value of the
solid surface value for all possible orientations, which by the same
arguments applied to (4.7) is clearly not true. If all possible orientations
of the surface of a solid can melt without nucleation barrier then

$$\{\gamma_S(hkl)\}_{min} > \gamma_L + \gamma_{SL}. \tag{4.10}$$

However, the error is once again probably no worse than 10 per cent
or so, and may account for the consistently low values noticeable (table
4.1). In this case the theoretical values do represent a form of average
of γ_{SL} over the probable orientations of the solid and so provide a
suitable comparison for the results obtained from nucleation experiments.

One interesting feature which would be worthy of theoretical consideration is the anisotropy of γ_{SL} as a function of orientation. The
nearest neighbour bonding models are largely unsuitable for discussing
such a problem but Zadumkin specifically anticipates an anisotropy of
exactly the same degree as for the free solid surface. Experimental
results are lacking for pure materials but the results of measurements
on alloys obtained by Miller & Chadwick (1969) seem to indicate that
this is not the case. For example, cubic materials seem to be entirely
isotropic over their solid–liquid interface although the same materials
are known to have a cusped γ_S-plot.

In fact the apparent success of broken bond models is not really very
significant. For example, if we choose a nearest neighbour bonding
model, then on the assumption that the difference in internal energy
between an atom in its solid and liquid states is equally shared between
its nearest neighbour bonds, then the surface energy per atom on the
surface will be $(\eta_0/\nu)L_0$ in the notation of the Jackson model (see §3.2).
For a general (high index) orientation surface there will be no nearest
neighbour bonds parallel to the surface and so $\nu = 2\eta_0$ and the surface
energy is $0.5L_0$ per atom site. Turnbull's experiments give a value of
$0.45L_0$ per atom site. For lower index orientations, (η_0/ν) will fall below
0.5 and it is clear that adding some longer range force interactions will
have the effect of reducing the marked anisotropy between these
orientations and will slightly reduce the energy of the general high index
orientation surface. In other words, even quite a cursory look at the
theory yields apparently quite good agreement.

However, in all these calculations it is the surface (internal) energy which has been calculated and compared with the experimental surface *free* energy. No account of surface entropy has been included in the theory. This can be done quite simply by using the equilibrium configurations determined by the Jackson model (or Temkin's more extensive theory). Unfortunately, this has the result of reducing the theoretical values considerably. Indeed, particularly for low index orientations on low γ-factor materials the resulting free energy can be negative at the melting point which is clearly nonsense.† The Temkin results suffer from the same problem. The reason for low theoretical free energies at low index orientations can be attributed to the low internal energies resulting from the nearest neighbour approximation as seen in the remarks above. Apparently, this approximation also has the effect of also seriously overestimating the surface entropy contribution. Unfortunately, in the severe absence of experimental results on surface entropies and anisotropy of free energy (and far from an excess of any form of free energy results), it is not clear how the theories can best be developed further.

Further reading
General discussions of most of the content of this chapter are not available elsewhere and so the reader wishing to pursue any point in detail will have to refer to the original papers cited. Some specific areas that are of fringe interest are dealt with elsewhere; in particular a review of phenomenological theories of melting is given by Ubbelohde (1965).

† This point has been the subject of some recent correspondence in the literature. Nason and Tiller have questioned the validity of Jackson's treatment in view of this inconsistency (Nason & Tiller, 1971; Jackson, 1971).

5 Morphological stability

5.1 Introduction

In all the discussions so far we have assumed that the morphology of an interface is determined either at equilibrium by the anisotropy of the interfacial free energy when the two phases are in equilibrium, or by the degree of roughness or diffuseness of the interface if growth of one of the phases is occurring. This would imply that during the solidification of a metal, for example, the interface being rough would always follow an isotherm of the system. In uniaxial growth this would lead to a planar interface, and in a bulk casting, cooled sufficiently slowly to avoid large temperature gradients until nucleation occurred at many sites in the liquid, spherical grains would grow into one another due to the radial temperature fields set up around the nuclei as the latent heat is given up. However, experiment shows that this is not the case and that these morphologies are not stable. Unless very special conditions are imposed, metals always freeze with a dendritic (tree-like) structure. We shall discuss the particular form of the dendrite and the origins of dendritic growth in the next chapter. In the present chapter we look at the general principles involved in determining the stability of an interface morphology by firstly considering a particularly simple case; this is the stability of a planar interface advancing at a fixed velocity under steady-state conditions in a binary alloy system. The discussion will largely neglect the effects of anisotropy of growth rates and faceting. In fact the marked preference of a faceting material to develop the simple growth facet tends to make this morphology stable under conditions which would produce considerable instability in non-faceting materials.

In the next chapter the morphology of dendrites will be analysed in more detail. This involves first of all considering the stability of a small spherical nucleus to determine why the spherical shape is not maintained, and then examining the stability of a spike to see why secondary arms appear on this. Finally, the dendritic shape itself will be investigated to try to understand why this growth form is then stable.

A qualitative understanding of the most common growth form of dendrites can be achieved simply by considering what happens when a pure metal melt solidifies. First the temperature must be reduced to some value below the melting point in order that the solid phase may be nucleated (this supercooling is very large if nucleation occurs

homogeneously but must still be finite even if the mechanism is heterogeneous). We now have a small nucleus of solid, which for simplicity
we will assume to be spherical, growing into the melt which is at a
temperature below the bulk melting point. The spherical symmetry implies that the temperature fields will be radial and the interface follows
an isotherm. Now as the heat source in the system is latent heat from the
interface, and the sink is outside the casting, it is clear that a negative
temperature gradient exists around the nucleus. Consider now the effect
of a local perturbation on the interface between the nucleus and melt
disturbing it slightly from its spherical shape. Those parts of the interface furthest from the centre of the nucleus will be in a region of greater
supercooling than those nearest the centre and so will grow faster. In
other words, the amplitude of this perturbation will *increase* with time.
The spherical shape is therefore an unstable morphology in that the
ambient conditions amplify any small disturbance of the system. We
see therefore how spikes can grow out of the sphere and hence how the
elements of the dendritic morphology are produced. Note, however,
that the crystallographic symmetry of the dendrite has not been explained; we will return to this point in the next chapter.

In order to understand more clearly the origins of interface instability
we shall now return to the problem of steady-state growth of a binary
alloy under planar interface conditions. While an alloy is more complicated theoretically than a pure material, it is more suitable for experimental studies due to the greater control which can be maintained (see
§5.3). Moreover we shall see that the theoretical problem need not be
made significantly more difficult by this.

5.2   Steady-state growth conditions for a binary alloy

In order to investigate the stability of a particular morphology, the
forms of the heat and solute diffusion fields of the unperturbed situation
must first be determined. We therefore consider the uniaxial solidification of a binary alloy having a partition coefficient† k which for convenience is assumed to be independent of alloy composition in the range
of interest. The relevant part of the phase diagram is therefore shown in
fig. 5.1. In this diagram and subsequent ones k is assumed to be less than
unity; this is not a necessary assumption of the arguments, however,

---

† The partition coefficient is the ratio of the liquidus and solidus gradients in
the phase diagram. An elementary knowledge of phase diagrams is assumed
but can otherwise be quickly attained from any standard metallurgical
text; e.g. Chalmers (1959), Cottrell (1967) or the text on phase diagrams by
Rhines (1956).

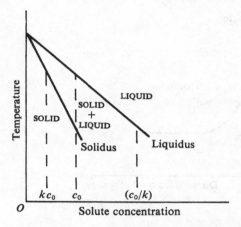

Fig. 5.1. Part of a binary alloy phase diagram with partition coefficient k less than 1.

but it will be convenient to discuss the effects in relation to this case.

Now consider the uniaxial solidification of a melt with an average solute concentration c_0 in which the temperatures are maintained such that the interface advances at a constant velocity V. The phase diagram shows that the first solid to form will have a solute concentration kc_0 to be in equilibrium with the liquid of concentration c_0. Now as the interface advances into the melt, the solid so formed will be purer than the liquid it replaces and hence solute will be pushed ahead of the interface. This, in turn, will of course increase the concentration of solute in the liquid at the interface and hence further freezing must occur at a slightly lower temperature and richer composition. Thus the interface equilibrium conditions will 'slide down' the solidus and liquidus until the solid is being formed with a concentration of solute c_0. This must represent the limiting 'steady-state' condition; if the limiting value were less than c_0, a net increase of solute would occur in time on the melt side of the interface and similarly, for a value greater than c_0 there would be an increase on the solid side. Thus, after some 'incubation time', or distance, a steady state will be reached when the composition of the solid freezing out has a constant solute concentration c_0, while ahead of the interface there will be a pile-up of solute for which the rate of rejection of solute from the solid phase exactly equals the diffusive flux away from the boundary into the liquid. Thus, the composition profile remains constant relative to the interface. This steady-state configuration is shown schematically in fig. 5.2. To determine its form analytically

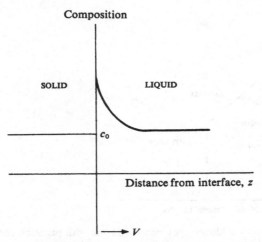

Fig. 5.2. Variation of composition with distance from the interface in steady-state solidification at a rate V.

we must consider the diffusion equation for the solute. In one dimension this is simply

$$D \frac{\partial^2 c}{\partial z'^2} = \frac{\partial c}{\partial t} \tag{5.1}$$

where D is the diffusion coefficient of the solute in the phase considered. In fact we can neglect effects in the solid not only because D for the solid phase is small compared with that for the liquid, but also in the steady state $(\partial c/\partial z')$ is zero in the solid. Also in the steady state (by definition), $(\partial c/\partial t) = 0$ in the frame of reference of the interface. Thus changing to this frame (denoted by z, the distance from the interface) we have

$$D_L \frac{\partial^2 c_L}{\partial z^2} + V \frac{\partial c_L}{\partial z} = 0 \tag{5.2}$$

where V is the velocity of propagation of the interface. Now integrating this expression, and applying the boundary conditions that for $z = 0$, $c_L = c_0/k$ and at $z \to \infty$, $c_L \to c_0$ (and hence $(\partial c/\partial z) \to 0$) we arrive at the solution

$$c_L = c_0 \left\{ 1 + \frac{(1 - k)}{k} \exp\left(-\frac{Vz}{D_L}\right) \right\}. \tag{5.3}$$

5.3   Constitutional supercooling
In order to look at the interface equilibrium in detail it is necessary also to know the form of the thermal diffusion fields near the interface.

This can be done by solving equations like (5.2) for the temperature in the liquid (T_L) and solid (T_S) as follows:

$$D_{Lt}\frac{\partial^2 T_L}{\partial z^2} + V\frac{\partial T_L}{\partial z} = 0$$

and

$$D_{St}\frac{\partial^2 T_S}{\partial z^2} + V\frac{\partial T_S}{\partial z} = 0 \tag{5.4}$$

where D_{Lt} and D_{St} are the thermal diffusivities of the liquid and solid phases which are related to the thermal conductivities K_L and K_S by

$$D_{St} = \frac{K_S}{\rho_S s_S} \quad \text{and} \quad D_{Lt} = \frac{K_L}{\rho_L s_L} \tag{5.5}$$

where the ρs are the densities and the ss the specific heats of the two phases. Solving these with the boundary condition that at the interface ($z = 0$) the temperature $T_L = T_S = T_i$, say, and that the temperature gradients in the two phases are given by G_L and G_S near the interface, we have

$$T_L = T_i + \frac{D_{Lt}G_L}{V}\{1 - \exp(-zV/D_{Lt})\},$$

$$T_S = T_i + \frac{D_{St}G_S}{V}\{1 - \exp(-zV/D_{St})\}. \tag{5.6}$$

A major simplification can be applied to these equations because we will only be interested in the temperature distribution in the region in which the solute concentration is varying markedly. Thus we are interested in the region $zV \leqslant D_L$. D_L will typically be of the order of 10^{-5} cm^2 s^{-1}, whereas D_{St} and D_{Lt} have magnitudes of the order of unity in the same units. Thus within this region $zV \ll D_{St} \sim D_{Lt}$ and hence (5.6) will simplify to

$$T_L = T_i + G_L z,$$

$$T_S = T_i + G_S z. \tag{5.7}$$

That is, in the region of interest the temperature gradients can be assumed to be linear. Moreover, the relative values of the gradients in the two phases can be determined by considering the heat transfer and generation of latent heat at the interface which leads to the equation

$$VL = K_S G_S - K_L G_L \tag{5.8}$$

where L is the latent heat of fusion of the material.

To consider the stability of the situation we now compare the actual distribution of temperature in the system with the variation in freezing temperature (i.e. the liquidus temperature). The liquidus temperature is directly related by means of the phase diagram to the solute distribution determined in (5.3). Fig. 5.3 shows the result of this. For the time being

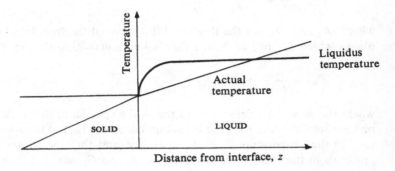

Fig. 5.3. Local freezing temperature-and actual temperature variation near a steady-state solidification interface as in fig. 5.2.

we assume that the interface temperature T_i is also the equilibrium temperature for the solid and liquid adjacent to the boundary. This therefore neglects any kinetic effects (i.e. the small supercooling necessary to 'drive' the solidification). The essential feature is that there may be a region in the liquid in which the actual temperature is lower than the liquidus temperature. This corresponds to a region of supercooling, and if solid were present here it would grow rapidly. (Nevertheless, the interface itself is at equilibrium in that the actual temperature and the liquidus temperature are the same there; this was one of the applied boundary conditions.) Moreover, if this region of supercooling exists it is such that the degree of supercooling increases as we move away from the interface. The situation is therefore similar to the case of a pure material growing into a negative temperature gradient as described in §5.1 and the planar interface shape should be unstable for the same reason. If any part of the interface is perturbed to move ahead of the main interface it enters a region of increasing supercooling and thus it grows even more rapidly. Thus the amplitude of the perturbation increases and therefore the interface is unstable. The condition giving rise to this effect is known as *constitutional supercooling*, for obvious reasons. The condition for it to occur is that G_L shall be less than the liquidus gradient in the liquid at the boundary. Thus if

$$T_{\text{liquidus}} = T_m + mc_L$$

where T_m is the pure solvent melting point and m is the liquidus slope, the condition is given by

$$G_L < mc_0 \frac{(k-1)}{k} \frac{V}{D_L} = mG_{cs}. \qquad (5.9)$$

It is found experimentally that this instability does indeed occur. However a new interface morphology evolves which then becomes stable within certain limits. The new morphology is referred to as *cellular* because a transverse section of a solid grown with this interface has a honeycomb or cellular structure. A typical cellular growth interface is seen in fig. 5.4. Fig. 5.5 shows a decanted interface viewed along the growth direction. This effect was first explained by Rutter & Chalmers (1953) and discussed in the quantitative terms put forward here shortly afterwards by Tiller, Jackson, Rutter & Chalmers (1953). The fact that the cellular interface is itself stable, at least for reasonable amounts of constitutional supercooling, is attributed to the fact that the new morphology reduces the constitutional supercooling. It does this by depositing solute-rich liquid in the grooves between the cells which then freezes well back from the main interface at the lower temperature to be found there; the build-up of solute ahead of the main interface is thereby reduced and thus the constitutional supercooling is also reduced.

A binary alloy affords a far more convenient system for experimental study of instability in a planar interface morphology than that provided by a pure solid–melt system. This is because in the case of the pure material (once again considering only metal-like growth) growth is almost entirely limited by heat transfer. In order to achieve the negative temperature gradient necessary for instability, the rate of heat removal must be very rapid and hence the growth is also rapid; under these conditions it is difficult to measure the important parameters affecting growth with any accuracy. In the case of a binary alloy, however, the growth will be much slower and more controlled, being limited by the diffusion of solute away from the interface rather than heat. It is therefore easier to maintain steady-state conditions for quite long periods of time and so ensure that the experimental conditions are comparable with the theoretical situation discussed here. One further interesting aspect of this system is that the transfer from one stable morphology to another (planar to cellular) offers an even more definitive test of theory. If the new morphology is well defined it might even be possible to predict the new shape exactly.

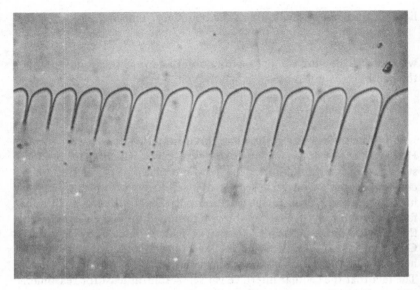

Fig. 5.4. A cellular interface growing in carbon tetrabromide, × 140 (courtesy Dr J. D. Hunt).

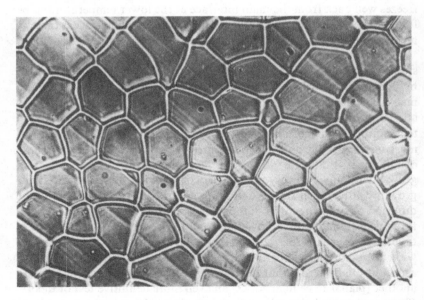

Fig. 5.5. Decanted cellular interface in aluminium crystal, × 1000 (courtesy Dr A. Hellawell).

5.4   Mullins and Sekerka's analysis of planar interface stability

We shall now investigate the stability criterion for the planar to cellular transition in a rather more rigorous fashion. The above arguments neglect, in particular, two possible stabilising effects on any interface perturbation; these are diffusion of solute and heat around the perturbation, and the effect of the finite interfacial free energy which will try to keep the interface planar in order to reduce the interfacial area.

The approach to the problem is that used by Mullins & Sekerka (1964). We consider a general perturbation on the interface shape and investigate the time-dependence of the amplitude of the perturbation due to the effects of diffusion and surface tension. In particular, as any general function can be represented by a Fourier series of sinusoidal functions, we need only consider sinusoidal perturbations of all possible wavelengths. Then if the rate of growth of the perturbation is negative for all wavelengths, no form of perturbation will grow and so the interface must be stable. On the other hand if the rate of growth is positive for any value or range of values of the wavelength, then the interface is unstable though the exact morphologies into which it will transform must be the subject of further analysis. It is worth noting that the perturbation can alternatively be applied to the local diffusion fields to give the same results.

The case to be considered is that of steady-state solidification already discussed in §5.2. To simplify the analysis it will be assumed that all bulk and surface parameters are isotropic, and that there is local equilibrium at the interface. We have, of course, already shown that the interfacial free energy is probably not isotropic, but this is a useful initial simplification and may not be too bad for many metals. The condition of local equilibrium really means that all possible kinetic effects are neglected. One kinetic effect which will be present in a real growth situation is that the interface temperature will differ from the equilibrium one by a small amount of supercooling, acting as the driving force for solidification and being a function of the rate of growth. This point will be allowed for later but we must assume that there are no other effects; certainly if there are they are not understood. A more fundamental simplifying assumption is the use of steady-state thermal and diffusion fields (i.e. non-time-dependent diffusion equations) when perturbations whose amplitudes are time dependent are imposed. This is evidently not a strictly valid assumption but the real situation will approximate to this as the rate of growth of perturbations tends to zero. That is, if the perturbations are sufficiently slow growing a quasi-steady state will prevail. In fact this approximation has been subjected to scrutiny by investigating the proper time-dependent case. Sekerka (1967*a, b*) has done this using a Laplace transform in time to achieve a solution. He concludes that if the thermal steady state is a good approximation, the stability criteria are the same as for the completely steady state model. As has been pointed out, the rates of thermal diffusion are very rapid so that this is likely to be a good approximation. Moreover, as Mullins and Sekerka have pointed out, the steady-state analysis, being valid for sufficiently slowly varying perturbations, should correctly predict the stability criterion. This is because if the theory predicts an unstable interface, in that there is a range of values of the wavelength and hence ω for which the rate of change in amplitude per unit amplitude $(\dot{\delta}/\delta)$ is positive, then there must be a narrow range of wavelengths for which $(\dot{\delta}/\delta)$ is infinitesimally small and hence satisfies the steady-state condition. Thus these alone will ensure instability within the framework of the model.

To proceed with the analysis we therefore apply a sinusoidal ripple of very small amplitude δ to the interface, described by

$$z = \phi(x, t) = \delta(t) \sin x. \tag{5.10}$$

Notice that only a two-dimensional model is adopted as it is assumed that this will simplify the analysis while in no way affecting the final results. Now we must solve the basic steady-state diffusion equations for solute diffusion in the liquid, and heat diffusion in both phases ((5.2) and (5.4)). It is assumed that solute diffusion in the solid will not contribute significantly due to the small values of solid diffusion coefficients.

Now in solving the diffusion equations we can apply various boundary conditions. The first of these is given by conservation of heat and solute concentration across the interface (cf. (5.8)):

$$v(x) = \frac{1}{L} \left\{ K_s \left(\frac{\partial T_s}{\partial z} \right)_\phi - K_L \left(\frac{\partial T_L}{\partial z} \right)_\phi \right\}$$

$$= \frac{1}{c_{L\phi}} (k - 1) D_L \left(\frac{\partial c_L}{\partial z} \right)_\phi \tag{5.11}$$

and also the conditions for equilibrium at the interface. Now the equilibrium temperature of the interface will be the pure solvent melting temperature T_m, modified by an amount $mC_{L\phi}$, by the presence of a concentration $c_{L\phi}$ of solute, and also by a capillarity term due to the curvature of the interface at the perturbation. The capillarity effect has already been discussed in chapter 2; if the radius of curvature (centred in the solid phase) is r, this is $(-T_m\gamma/rL)$. Therefore writing $\Gamma = \gamma/L$ and $1/r = \delta\omega^2 \sin \omega x$, gives the boundary condition

$$T_\phi = T_m + mc_{L\phi} - T_m\Gamma\delta\omega^2 \sin \omega x. \tag{5.12}$$

Now if the perturbation amplitude δ is small, linear solutions of the form

$$T_\phi = T_0 + a\delta \sin \omega x = T_0 + a\phi$$

and
$$c_{L\phi} = c_{L0} + b\delta \sin \omega x = c_{L0} + b\phi \tag{5.13}$$

can be selected, a and b being coefficients to be determined. This leads to the solutions as follows:

$$c_L(x, z) - c_{L0} = (G_{cs}D_L/V)[1 - \exp(-Vz/D_L)]$$
$$+ \delta(b - G_{cs}) \sin \omega x \exp(-\omega_L^* z) \tag{5.14}$$

$$T_L(x, z) - T_0 = (G_L D_{Lt}/V)[1 - \exp(-Vz/D_{Lt})]$$
$$+ \delta(a - G_L) \sin \omega x \exp(-\omega_{Lt} z) \tag{5.15}$$

$$T_S(x, z) - T_0 = (G_S D_{St}/V)[1 - \exp(-Vz/D_{St})]$$
$$+ \delta(a - G_S) \sin \omega x \exp(-\omega_{St} z) \tag{5.16}$$

where
$$\left. \begin{aligned} \omega_L^* &= (V/2D_L) + [(V/2D_L)^2 + \omega^2]^{1/2} \\ \omega_{Lt} &= (V/2D_{Lt}) + [(V/2D_{Lt})^2 + \omega^2]^{1/2} \\ \omega_{St} &= (V/2D_{St}) - [(V/2D_{St})^2 + \omega^2]^{1/2}. \end{aligned} \right\} \tag{5.17}$$

The expressions (5.17) are solutions of quadratic equations, and the. choice of sign is such that $\exp(-\omega_{St} z)$, $\exp(-\omega_L^* z)$ and $\exp(-\omega_{Lt} z)$ decay as $|z|$ increases into the phase to which they relate. Thus equations (5.14) to (5.16) simplify to the unperturbed steady-state values a long way from the interface.

Now (5.12) and (5.13) show that

$$a = mb - T_m \Gamma \omega^2 \tag{5.18}$$

and to separate a and b, the gradients of the diffusion fields can be evaluated from equations (5.14) to (5.17) as follows:

$$(\partial c_L/\partial z)_\phi = -\omega_L^*\{b - G_{cs}[1 - (V/\omega_L^* D_L)]\}\delta \sin \omega x + G_{cs} \tag{5.19}$$

$$(\partial T_L/\partial z)_\phi \simeq -\omega(a - G_L)\delta \sin \omega x + G_L \tag{5.20}$$

$$(\partial T_S/\partial z)_\phi \simeq +\omega(a - G_S)\delta \sin \omega x + G_S \tag{5.21}$$

where the approximations involved in (5.20) and (5.21) are that $D_{Lt}\omega \gg V$ and $D_{St}\omega \gg V$ for all practical values of these parameters so that $\omega_{\text{Lt}} \simeq \omega \simeq -\omega_{\text{St}}$. Then substituting these gradients into the boundary conditions (5.11) allows b to be evaluated as

$$b = \frac{2G_{\text{cs}}T_{\text{m}}\Gamma\omega^3 + \omega G_{\text{cs}}(\mathscr{G}_{\text{S}} + \mathscr{G}_{\text{L}}) + G_{\text{cs}}[\omega_{\text{L}}^* - (V/D_{\text{L}})](\mathscr{G}_{\text{S}} - \mathscr{G}_{\text{L}})}{2\omega m G_{\text{cs}} + (\mathscr{G}_{\text{S}} - \mathscr{G}_{\text{L}})[\omega_{\text{L}}^* - (V/D_{\text{L}})p]} \quad (5.22)$$

where $\mathscr{G}_{\text{L}} = (K_{\text{L}}/\bar{K})G_{\text{L}}$, $\mathscr{G}_{\text{S}} = (K_{\text{S}}/\bar{K})G_{\text{S}}$, $\bar{K} = \frac{1}{2}(K_{\text{S}} + K_{\text{L}})$ and $p = (1 - k)$. Now to determine $\dot{\delta}$, we substitute (5.20) and (5.21) into the first part of (5.11) to get

$$v(x) = V + \dot{\delta} \sin \omega x$$
$$= (\bar{K}/L)\{(\mathscr{G}_{\text{S}} - \mathscr{G}_{\text{L}}) + \omega[2a - (\mathscr{G}_{\text{S}} + \mathscr{G}_{\text{L}})]\delta \sin \omega x\} \quad (5.23)$$

and equating coefficients in this equation gives

$$V = (\bar{K}/L)(\mathscr{G}_{\text{S}} - \mathscr{G}_{\text{L}})$$

and

$$\dot{\delta} = (2\bar{K}/L)\omega[a - \frac{1}{2}(\mathscr{G}_{\text{S}} + \mathscr{G}_{\text{L}})]\delta. \quad (5.24) \text{ (cf. (5.8))}$$

Hence, substituting for a from (5.18) and (5.22) leads to the central result

$$\frac{\dot{\delta}}{\delta} = \frac{V\{-2T_{\text{m}}\Gamma\omega^2[\omega_{\text{L}}^* - (V/D_{\text{L}})p] - (\mathscr{G}_{\text{S}} + \mathscr{G}_{\text{L}}) \times [\omega_{\text{L}}^* - (V/D_{\text{L}})p] + 2mG_{\text{cs}}[\omega_{\text{L}}^* - (V/D_{\text{L}})]\}}{(\mathscr{G}_{\text{S}} - \mathscr{G}_{\text{L}})[\omega_{\text{L}}^* - (V/D_{\text{L}})p] + 2\omega m G_{\text{cs}}}. \quad (5.25)$$

This expression tells us about the time evolution of a perturbation of wavelength $(2\pi/\omega)$ within the region of applicability of the model used (this particularly means that it is only meaningful for very small amplitude perturbations). In order for the planar interface morphology to be stable we must have $(\dot{\delta}/\delta)$ negative for all ω; if $(\dot{\delta}/\delta)$ is positive for any ω, then perturbations having a corresponding wavelength will grow and change the morphology (fig. 5.6). In fact we need only concern ourselves with the sign of the numerator of (5.25) as the denominator is always positive; considering each part of the denominator, $(\mathscr{G}_{\text{S}} - \mathscr{G}_{\text{L}})$ is proportional to V and therefore positive (equation (5.24)), $[\omega_{\text{L}}^* - (V/D_{\text{L}})p]$ is positive because $\omega_{\text{L}}^* \geqslant (V/D_{\text{L}})$ (equation (5.17)) and hence $\omega_{\text{L}}^* \geqslant (V/D_{\text{L}})p$ as $p = (1 - k) \leqslant 1$, and finally $2\omega m G_{\text{cs}}$ is positive because if $k < 1$, m and G_{cs} are both negative whereas if $k > 1$, m and G_{cs} are both positive. Thus dividing the numerator through by the

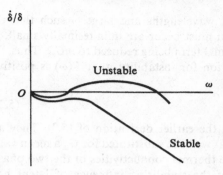

Fig. 5.6. Stable and unstable forms of possible $(\dot{\delta}/\delta)$ relationships with ω.

positive quantity $2[\omega_L^* - (V/D_L)p]V$ leads to the function $S(\omega)$ which must always be negative for stability

$$S(\omega) = -T_m \Gamma \omega^2 - \tfrac{1}{2}(\mathscr{G}_S + \mathscr{G}_L) + \frac{mG_{cs}[\omega_L^* - (V/D_L)]}{[\omega_L^* - (V/D_L)p]}.$$
(5.26)

Now, before proceeding to deduce the exact stability criterion given by this function, it is worth pausing to examine its physical significance. The first term, involving the capillarity constant has a stabilising influence for all wavelengths, though its effect is most favourable at short wavelengths (large ω). This is exactly the sort of stabilising effect we would expect of surface tension. The second term is also stabilising for positive temperature gradients; if the gradients are negative, the term is destabilising which is in accord with previous ideas as this represents the growth of a solid into a positive gradient of supercooling (indeed, for a pure material this is the only destabilising term). The final term is always destabilising, being positive, and represents the effect of solute diffusion.

It is interesting also to try to get back to the simple constitutional supercooling criterion derived earlier. To do this it is necessary to introduce the same simplifications implicit in that treatment. These were firstly that there was no surface tension effect, which can be achieved by putting $\Gamma = 0$ and hence removing the first term, and secondly that solute diffusion effects were such as to ensure that when one part of the interface grew ahead of its surrounding parts, these adjacent regions were not slowed down by the extra solute pile-up from the outwardly perturbed regions. This effectively means that we must assume solute diffusion which is perfectly efficient in displacing excess solute *across* the interface; this can be simulated, either by letting D_L increase without limit, or going to very short wavelengths and large ω such that the distances over which diffusion must occur are infinitesimally small. In either case this results in the final term being reduced to mG_{cs}. Thus, for this simplified case the criterion for instability, that $S(\omega)$ is positive, becomes

$$mG_{cs} > \tfrac{1}{2}(\mathscr{G}_S + \mathscr{G}_L). \tag{5.27}$$

This is to be compared with the earlier derivation of (5.9). They are essentially the same except that we have substituted for G_L a mean value of G_L and G_S weighted by the thermal conductivities of the two phases (which also takes account of the stabilising influence of latent heat generation via the boundary conditions (5.8) restated in (5.24)).

It is now clear that the fuller stability criterion will indicate a greater region of stability than (5.9) or (5.27). At the short wavelength limit used to derive (5.27) the neglected capillarity term has its greatest stabilising influence. At long wavelengths, where the capillarity term becomes unimportant, the final term tends to zero so that the only remaining term is the stabilising temperature gradient one. Indeed, in the long wavelength limit solute diffusion no longer destabilises the interface; in other words the diffusion of solute across the interface is the essential process causing instability.

Now, returning to the exact stability condition defined by $S(\omega) < 0$, this can be written in a slightly different form, separating out the wavelength dependent and independent parts of $S(\omega)$ by defining

$$G(\omega) = \frac{-T_m \Gamma \omega^2}{mG_{cs}} + F(\omega) \tag{5.28}$$

where

$$F(\omega) = \frac{[\omega_L^* - (V/D_L)]}{[\omega_L^* - (V/D_L)p]}. \tag{5.29}$$

The condition for *stability* now becomes

$$\tfrac{1}{2}(\mathscr{G}_S + \mathscr{G}_L) > mG_{cs}G(\omega). \tag{5.30}$$

$G(\omega)$ is composed of two parts, one proportional to ω^2 and the other part $F(\omega)$ being proportional to ω^2 for small ω and tending to unity for large ω. Fig. 5.7 shows the form of these two parts. The diagram also shows the two, basically different, possible forms of their sum $G_1(\omega)$ and $G_2(\omega)$. For small ω

$$F(\omega) = (D_L^2/V^{2k})\omega^2 \tag{5.31}$$

and it follows therefore that if the constant (D_L^2/V^2k) is less than the

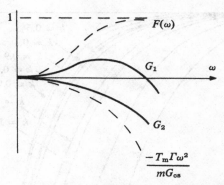

Fig. 5.7. The form of the two components of the function $G(\omega)$ and two possible forms of the sum.

constant in the capillarity term which is also proportional to ω^2, then G will have the form of the G_2 curve in fig. 5.7; that is, $G(\omega)_{\max} = 0$. In this case we have a situation of '*absolute stability*'; providing the thermal gradients are not actually negative the interface must be stable even though these gradients may be much less than the liquidus temperature gradient mG_{cs}. This absolute stability condition is therefore

$$\frac{T_{\mathrm{m}}\Gamma}{mG_{\mathrm{cs}}} > \frac{D_{\mathrm{L}}^2}{V^2 k} \quad \text{or} \quad A > 1 \tag{5.32}$$

where

$$A = \frac{kT_{\mathrm{m}}\Gamma V^2}{mG_{\mathrm{cs}}D_{\mathrm{L}}^2} = \frac{k^2 T_{\mathrm{m}}\Gamma V}{(k-1)mc_0 D_{\mathrm{L}}}. \tag{5.33}$$

If, on the other hand, this inequality (5.32) is not obeyed, $G(\omega)$ will have a maximum for some ω greater than zero like G_1 in fig. 5.7. Bearing in mind that stability can exist only when (5.30) is satisfied for all ω, the general condition for stability must be given by

$$\tfrac{1}{2}(\mathscr{G}_{\mathrm{S}} + \mathscr{G}_{\mathrm{L}}) > mG_{\mathrm{cs}}\mathscr{S} \tag{5.34}$$

where

$$\mathscr{S} = G(\omega)_{\max}. \tag{5.35}$$

It is worth noting that in this form the comparison with the constitutional supercooling criterion (5.27) is particularly clear as this corresponds to $\mathscr{S} = 1$. In fact $\mathscr{S}$ has a maximum possible value of unity and will in general be lower (and can be zero in the case of absolute stability defined by (5.32)). The problem of tabulating the function $\mathscr{S}$ has been tackled by Sekerka (1965) and his results are shown in fig. 5.8 in which curves of $\mathscr{S}$ versus the parameter A are given

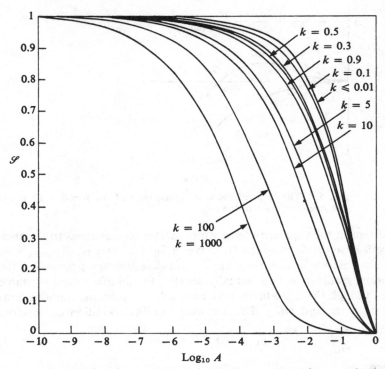

Fig. 5.8. Plot of the stability function $\mathscr{S}$ versus log A for several values of k. (After Sekerka, 1965.)

for various values of the partition coefficient k, of the system. This permits the exact stability criterion for any particular system to be determined. Indeed, it is possible to recast (5.34) in a clearer form by defining a 'gradient of constitutional supercooling', G', which is equal to the difference between the liquidus and actual temperature gradients at the interface, $(mG_{cs} - G_L)$. In the form the theory predicts stability when

$$G' < \frac{2K_L}{K_S + K_L}\left(\frac{LV}{2K_L} + G_L\right)\frac{1}{\mathscr{S}} - G_L. \qquad (5.36)$$

The exact condition now becomes clear; the normal simplified constitutional supercooling criterion involves neglecting the stabilising influence of surface tension and latent heat generation. These approximations can be achieved above by putting $L \to 0$ and $\mathscr{S} \to 1$ when the condition becomes $G' < 0$; i.e. instability occurs when $G' > 0$ or when there is constitutional supercooling. In reality, L is positive and $\mathscr{S} < 1$, both of which increase the right-hand side of (5.36). Thus constitutional supercooling is a necessary but not sufficient condition for instability; the degree or gradient of constitutional supercooling must exceed some specified value. One possibility that is neglected in this discussion is that $K_S \gg K_L$ in which case it would be possible, were L also small, for (5.36) to predict instability with G' negative or without constitutional supercooling; in fact this condition is never likely to be achieved except at very large G_L (a condition which of course greatly reduces constitutional supercooling). Note also in (5.36) that in the limit of $\mathscr{S} \to 0$ (given, from fig. 5.8, by $A \geqslant 1$) the condition becomes $G' > \infty$; i.e. there is absolute stability as already defined.

5.5   The influence of interface kinetics on stability

In the analysis so far it has been assumed that the situation at the interface is one of equilibrium. This is obviously not true; one phase is growing at the expense of the other so the system must by definition deviate from equilibrium. This state is generally represented by considering the difference in temperature between the moving interface and a corresponding equilibrium one under otherwise similar conditions. For the case of solidification, therefore, there will be a finite interfacial undercooling providing the driving force for growth; moreover, this supercooling must be an increasing function of the growth rate (the greater the supercooling and hence driving force, the greater the growth rate). The particular form which this function takes is dependent on the material, and possibly also on the conditions of growth. Seidensticker (1967) has shown that this effect of interface kinetics can be allowed for quite simply and in a general way for the purposes of analysing interfacial stability. The method relies on the fact that for a sufficiently small region any function may be regarded as linear, and that in this analysis we are concerned only with very small variations in the growth rate (from $V + \dot{\delta}$ to $V - \dot{\delta}$ where $\dot{\delta} \ll V$) and hence also in the undercooling. Thus while the particular relation between interface undercooling and velocity of growth may be complex, in the region of interest for stability analysis it may be considered to be linear. We may write therefore

$$T_{\text{equilibrium}} - T_{\text{actual}} = \delta T_0 + \frac{1}{\mu}\phi \qquad (5.37)$$

where δT_0 represents the kinetic undercooling of the unperturbed interface and μ is some kinetic constant. It simply remains to include this in the thermal boundary condition (5.12) and to rework the analysis. This leads to a new stability equation for $(\dot{\delta}/\delta)$ which is identical to (5.25) except that the denominator has an additional term

$$+ \omega V/\mu[\omega_{\mathrm{L}}^* - (V/D_{\mathrm{L}})p].$$

(5.38)

This extra term is always positive, the denominator remains positive as for (5.25), and hence the stability criterion is determined by the sign of the unchanged numerator. Thus the inclusion of interface kinetics does not change the stability criterion. Actually, in a constrained system in which the external temperatures are controlled, it is possible that the change in interfacial undercooling will change G_{S} and G_{L} which can affect the numerator and therefore the stability. However, this is a rather superficial problem and is only a result of internal parameters changing while external ones remain fixed; it is not a fundamental change and does not alter the constitutional undercooling criterion which is explicitly a function of the thermal gradients.

The effect which the kinetics term does have is that for any particular value of ω, $|\dot{\delta}/\delta|$ is reduced; thus it reduces the *rate* of growth of perturbations. This can, in principle, be a very important effect for if the rate of growth is sufficiently small, it may result in a predicted instability never being of sufficient magnitude to be observed. Obviously it is not possible to be more specific without knowing the value of the parameter μ, and particularly the relative magnitudes of $(\mathscr{G}_{\mathrm{S}} - \mathscr{G}_{\mathrm{L}})$ and $(\omega V/\mu)$. μ is both a function of the particular system under consideration and also, more generally, the mechanism of solidification. In fact for most metal-like systems μ is sufficiently large for the kinetic undercooling term to be unimportant in these considerations. For the materials which grow with a singular interface, μ is much smaller and the kinetic term may dominate the denominator of the expression for $(\dot{\delta}/\delta)$ and therefore be the limiting factor controlling the rate of instability evolution. This is probably one of the reasons why such materials can be grown with such large and apparently stable facets under conditions of constitutional supercooling and even complete supercooling, when instability would be anticipated. Two other effects could also contribute to this, however. The first of these is that, as we shall see later, the growth on these facets is source limited; that is, growth can only originate from a finite number of sites on the interface, whereas this selectivity does not appear to occur on non-singular interfaces. If this is true then presumably for an instability of wavelength λ to grow there must be sufficient sources for them to be separated by a mean distance of less than λ; i.e. there must be at least one source per projection on the interface.

The second point relates to the obviously considerable anisotropy of either interfacial free energy, or interface kinetics, or both, which must exist in these materials. So far it has been assumed that both of these quantities are isotropic (although they are never likely to be exactly so as considerations in chapter 1 showed). The problem of putting these anisotropies into the analysis is a cumbersome one if nothing more. Of course, for small amplitude perturbations, the variation in orientations represented on the interface is small. However, as we have already seen, the unperturbed interfaces of interest in faceting materials are likely to be just the ones corresponding to a cusp in the γ-plot (and quite likely to a similar effect also in the kinetics). Thus, while the orientation changes considered are small, the interfacial free energies and kinetics may suffer quite large changes. In non-mathematical terms we might expect the interfacial free energy term, for instance, to produce a strong stabilising influence, favouring the observation of large facets under apparently unstable conditions. A more detailed treatment has been presented by Cahn (1967) which indicates that the principal reason for large facets is related to the kinetic effects and the particular mechanism by which growth is believed to occur in these materials. This will become clearer in the chapter on interface kinetics (chapter 8).

5.6   Time evolution of instabilities

One of the most serious approximations contained in the stability analysis presented above is the assumption that the diffusion fields can be represented by their steady-state values. Thus we are investigating a time-dependent effect by removing part of the time dependence from the problem. However, as we have pointed out, this should not be serious in determining the stable to unstable transition criterion because the time dependence is very weak near the critical condition. Indeed more thorough analyses, taking account of the time dependence on the diffusion fields, lead to almost identical results for the stability criterion.

The ideal theory would lead to two experimentally detectable and interesting predictions. The first of these is the point at which the system goes unstable. This we have already seen. The second would be to see *how* it goes unstable; in particular, what is the new shape which evolves from the instability. In fact, for the planar interface we already know that the shape evolving from the instability is the cellular structure (except in more exaggerated circumstances when the interface becomes dendritic). Therefore, it is interesting to see if our theory can predict the shape of the cellular interface. In fact this problem is formidable for reasons beyond the time-dependence limitations. These arise because there is a second major approximation involved in the theory as presented here in that the perturbations in the diffusion fields are linearly related to the perturbation amplitude on the interface; this is only true for very small amplitude perturbations (in particular, where the amplitude δ is very much less than the wavelength of perturbation λ). Obviously this condition does not apply to the cellular interface as observed experimentally and therefore the linear theories cannot be expected to predict the shape of the cellular interface.

No non-linear theory currently exists. Nevertheless, there is one way in which the observed cellular interface might be related to currently available theories. This is based on the assumption that the shape of the interface initially evolving from the planar morphology in the linear region will, to some extent, dictate the final observed shape. It might be expected (though it is certainly not obvious) that the wavelength corresponding to the fastest growing perturbation (i.e. the maximum value of $\dot{\delta}/\delta$) may dominate the instability and therefore correspond to the cell size that is observed experimentally in the new morphology.

It is at this point that we must reconsider the time-independent analysis. The validating assumption for obtaining the stability to instability criterion was that we only really had to believe the theory in the region where $(\dot{\delta}/\delta)$ was very small and positive. We now wish to use the *maximum* value of $(\dot{\delta}/\delta)$ to give further information about the shape of the new morphology emerging from the instability. To investigate this further Sekerka (1967b) has performed a proper time-dependent analysis but on the slightly simpler system of an isothermal phase transition which is limited by the diffusion in the matrix of one species. In this case he is able to investigate the time evolution of the interface and compare it with that predicted by the steady-state model (and also the 'Laplace equation' model – one which neglects the $V(\delta c_{\mathrm{L}}/\delta z)$ term in (5.2) and is used in the analyses of the stability of spheres and rods as discussed in the next chapter). For each of these models he compares two parameters, firstly the fastest growing wavelength perturbation and secondly the time constant τ associated with this growth rate. For the theory outlined above we have

$$\frac{\dot{\phi}}{\phi} = \frac{\dot{\delta}}{\delta}$$

and thus

$$\phi(\omega, t) = \phi_0(\omega) \exp\left(t(\dot{\delta}/\delta)\right) \tag{5.39}$$

which may be written as

$$\phi(\omega, t) = \phi_0(\omega) \exp\left(t/\tau\right). \tag{5.40}$$

He finds a marked variation between the predicted values of these parameters for the three models although they predict the same condition for the onset of instability. However, the theories are approximately in agreement on the time-evolution parameters for large values (i.e. $\gtrsim 1$) of

$$\eta = \frac{4\pi D_L}{V\lambda_0} \tag{5.41}$$

where λ_0 is the minimum wavelength for instability. Assuming that the same arguments apply to constitutional supercooling problems, rough values of η for typical experimental situations indicate that the steady-state theory should give reasonable values of the time-evolution parameters.

Sekerka also investigates in his paper the exact time evolution of some arbitrarily perturbed interface by taking the full Fourier transform to construct the actual interface shape as it develops with time. The calculations do indeed lead to an interface which is becoming sinusoidal with the wavelength corresponding to the maximum in $(\dot\delta/\delta)$ as we have supposed (which is independent of the initial form of the applied perturbation). However, Sekerka also points out that by the time this has been achieved the amplitude is well past the linear regime limit. Thus the need for caution in applying the theory mentioned earlier is borne out by more detailed calculations.

5.7   Experimental studies on constitutional supercooling

The breakdown of the planar interface into a cellular morphology in a uniaxially frozen binary alloy was first explained by Rutter & Chalmers in 1953. They proposed the constitutional supercooling criterion (§5.3) and showed that their experiments agreed with this. Subsequently many other experiments have led to similar conclusions. As the detailed stability theory outlined above leads to a very similar result we might well be inclined to ask why all this detailed analysis was thought to be necessary. The short answer is quite simply to put the constitutional supercooling theory on a sound footing. The original arguments were rather intuitive (though reasonable) in nature and it was not clear, for example, how much stabilising influence capillarity would exert on the interface. In fact the detailed analysis does predict certain differences from the results of the simple constitutional supercooling theory, but as yet these differences have not been shown up experimentally (although it is probably fair to say that no serious attempts have been made to do this). Most experiments on solidification have been performed under conditions where $A \ll 1$ and no marked difference is to be expected here. Furthermore, the approach adopted in many such experiments has been to deduce the values of physical parameters, such as D_L whose exact value is often not known, by assuming the constitutional super-cooling criterion, rather than trying to verify it. We should therefore ask whether the detailed theory can make any other predictions which can be tested experimentally. One possibility would be to compare the cell spacing with the wavelength of the most rapidly growing perturbation to which it may be related. We find that the absolute values of these quantities are indeed of the same order of magnitude. Experiment shows that the cell size is proportional to $V^{-1/2}$ but largely independent of temperature gradient. The theory predicts the velocity dependence correctly but there is less satisfactory agreement with the temperature gradient dependence (Sekerka, 1968). The problems of non-linearity have already been discussed so that this lack of correlation is certainly not unexpected. It does, however, underline the necessity for further experimental tests of very detailed theoretical studies. As Sekerka puts it, there is a need 'to keep the theory honest'. In fact there are some rather elegant experiments which have been performed and do to some extent offset this imbalance but discussion of these will be deferred to the next chapter. In addition, clear confirmation of the validity of stability theory of this type has been demonstrated for the melting case outlined in the following section.

Before leaving the case of the planar interface it may be worth describing some further experimental features which are of interest in understanding solidification effects but which can only be accounted for partially with simple 'first principles' theoretical models. If a binary alloy is solidified uniaxially we have already seen that at, or about, the critical condition for constitutional supercooling, the previously planar interface breaks up into a cellular one. These cells are initially normal to the main solidification front which roughly follows an isotherm of the system. If the constitutional supercooling is increased, however, the cells partially reorientate themselves towards the preferential directions of growth (corresponding to the normal dendritic growth directions mentioned in the next chapter), and with sufficient supercooling they grow side-arms and become dendritic.

It would obviously be of interest to find the conditions for this cell to dendrite transition. In fact the setting up of a model for this would be formidable. The shape of the cellular morphology cannot be predicted in detail, so some model based on experimental observations would have to be chosen; this in itself would be difficult as the depth of the cell boundaries is difficult to define. Indeed, in some systems at least, the cell boundaries can be so deep and so rich in solute that the ends neck off into liquid droplets (a non-steady-state behaviour which would be difficult to describe). Apart from such difficulties in describing the morphology there is some doubt as to whether this transition is in itself a well-defined one. Various experimenters have indicated that some condition related to a (G/V) or $(G/V^{1/2})$ parameter exists but a recent survey of the literature indicates that no real trend can be discerned in terms of any simple combination of measured quantities (Davies, 1968).

Finally, the experimental evidence for the occurrence of some structure on the interface prior to cell formation should be mentioned. Several workers have observed 'pox' on interfaces after decanting melts while the solid was being grown at conditions near the onset of constitutional supercooling. There is some variation in opinion as to the origin of these. In at least one case it has been suggested that the 'pox' are spikes on the decanted interface resulting from splashing during decanting or that they are depressions which are artefacts of the decanting technique (Spittle, Hunt & Smith, 1964). However, in other cases the pox were seen to be depressions and their association with dislocations in the solid has been suggested (Biloni, Bolling & Cole, 1966). These effects are clearly not predicted by stability theory but could be associated with the actual mechanism by which the interface goes unstable. Observations of the transition in transparent materials in thin optical cells shows that the instability usually starts at discernible singularities in the interface such as grain boundary cusps or specimen edges. From these points sinusoidal waves spread across the interface and grow into the cellular structure.

5.8   The stability of a melting interface

So far interface stability has been discussed only in connection with solidification. It is worth considering the relevance of these ideas to the process of melting. During the melting of a binary alloy it is clear that some sort of 'constitutional superheating' must be possible when the melting is controlled by diffusion of solute in the solid. First appearances would suggest that a similar instability should be associated with this. However, when a stability analysis similar to that described above (§5.4) is performed (Woodruff, 1968a) a condition is found which implies far greater stability than would be deduced from constitutional superheating criteria. The analysis involves including the diffusion equations for the solute in both phases because diffusion on either side of the interface now becomes important. This is because the diffusion flux is the product of two terms, the diffusion coefficient and the concentration gradient for 'steady-state melting'; the large concentration gradient in the solid counteracts the small coefficient, whereas in the liquid phase the small concentration gradient (zero for the unperturbed interface) is multiplied by the relatively large coefficient. (Note that in solidification both terms are large for the liquid and small for the solid so that diffusion in the solid can be safely neglected.) It is this similarity of diffusion fluxes in the case of melting which leads to the greatly enhanced stability condition. Constitutional superheating only takes account of the instability introduced by the concentration gradient term in the solid whereas in fact diffusion in both phases has a stabilising influence on the interface.

Recent experimental work has now provided confirmation of the main prediction of this analysis; that is, the greatly enhanced stability of the melting interface relative to the simple constitutional super-heating criterion. Verhoeven & Gibson (1971) have shown in melting experiments on Sn–Sb and Sn–Bi alloys that melting occurs with a stable planar interface well into the constitutional superheating regime, though proper stability is only obtained with sufficiently large temperature gradients. However, the breakdown of the melting pattern for low temperature gradients is thought to be due to nucleation of melting ahead of the interface and would not, therefore, be due to an instability of the interface itself. This work therefore provides the first clear evidence for the validity of stability theory results where they deviate from simpler (constitutional) considerations.

Further reading
The experimental aspects of the planar, cellular and dendritic solidification morphologies are covered in detail in Chalmers' (1964) book *The Principles of Solidification*. To gain more knowledge of the basic theory of morphological stability analyses the series of papers by Mullins and Sekerka, and by Sekerka are worthwhile reading, with quite a useful review of the field having been given by Sekerka (1968) in the Proceedings of the 1967 International Crystal Growth Conference. This review paper and Chalmers' book also provide extensive reference lists for those wishing to read even further into this subject.

6   Dendritic Growth

6.1   The origins and occurrence of dendritic growth

So far we have implied that the shape of the solid–liquid interface is largely controlled by the isotherms of the system or by its tendency to facet in growth. In the last chapter the idea of interfacial instability was introduced and shown to account in general terms for the cellular morphology, although a complete analysis of this proves to be difficult. Outside the controlled conditions of the laboratory, solidification almost exclusively occurs dendritically rather than with the relatively simple morphologies just described. The reason for this has already been outlined in §5.1. In the case of a pure melt, dendritic solidification occurs if there is sufficient supercooling of the melt so that this, together with the latent heat of solidification establishes the requisite negative temperature gradient. Outside the laboratory most melts are alloys and so in most practical cases constitutional supercooling effects will ensure the effective negative temperature gradient.

Dendritic growth is by no means restricted to growth from the melt. It can easily be seen in growth from solution or the vapour phase provided the supersaturation gradients are suitable. Indeed, instabilities of the dendritic type can be seen in situations quite unrelated to crystal growth; for instance, if a water drop is trapped between glass plates which are suddenly squeezed together the circular cross-section of the film becomes spikey as it grows and side arms may even be seen to develop. Even the branching of a tree may perhaps be thought of as an instability effect. (The name dendrite originates from the Greek word for a tree.) There is, however, one very important difference between these effects and crystal growth in that the directions of branching are essentially random. In crystals the dendrite directions are well defined in terms of the crystal axes and are quite reproducible. This particular anisotropy is of course to be expected, in that we know that, in general, crystal surface properties are anisotropic and reflect the symmetry of the crystal structure. This shows that any full description of dendritic solidification must also take account of the crystal anisotropy to explain these features. The characteristic growth directions for dendrites in the common crystal systems as reported by Chalmers (1964) are given in table 6.1. It turns out that this particular feature, which at first appearance might seem simple to explain, is in fact one of the main unresolved problems in the proper understanding of dendritic growth.

Table 6.1

Structure	Dendrite growth direction
f.c.c.	$\langle 100 \rangle$
b.c.c.	$\langle 100 \rangle$
h.c.p.	$\langle 10\bar{1}0 \rangle$
Body-centred tetragonal (tin)	$\langle 110 \rangle$

In attempting an account for dendritic growth two main approaches have been followed. Firstly (though historically most recently) attempts have been made to see when and how a spherical nucleus becomes unstable and grows spikes; then the analysis of the stability of a growing rod might lead to an understanding of branching of dendrite arms. Secondly, there have been attempts to explain how the observed morphology can be described in terms of the local flow of heat (and solute in alloys). There have also been many semi-empirical attempts to explain other features of dendrites, such as the spacing of arms as a function of growth rate. These complex but important problems will be discussed at the end of this chapter.

6.2   The stability of a growing spherical nucleus

The first attempt to analyse the stability of a growing sphere of solid from its melt was the subject of the original stability analysis paper of Mullins & Sekerka (1963). Subsequently Coriell & Parker (1967) and others have re-examined the problem introducing other effects such as interface kinetics. The formulation of the problem has been restricted to a precipitate of fixed concentration growing in a diffusion-controlled manner from a matrix under isothermal conditions. An associated problem of more interest to us is that of a pure solid growing from its own melt where growth is controlled by heat flow. As in the problem of the planar interface the basic approach is to apply a small perturbation to the initial shape of the interface and then look at the time evolution of the shape and in particular of the magnitude of the disturbance to the shape having the same wavelength as the applied perturbation. In this case a sphere of radius R is perturbed by some spherical harmonic $Y_{l,m}(\theta, \phi)$ having an infinitesimally small amplitude δ to give a solid–liquid interface defined by

$$r = R + \delta Y_{l,m}(\theta, \phi). \tag{6.1}$$

In this situation there can be no steady-state situation to perturb and so to analyse the stability effects in the simplest way we must use the time-independent form of the heat diffusion equation and require that the temperature in each phase, T_L and T_S, should satisfy Laplace's equation

$$\nabla^2 T_L = \nabla^2 T_S = 0. \tag{6.2}$$

The use of this simplifying assumption can be justified providing that the driving force is small so that the quasi-time-independent state is satisfied. Formally the condition in the case of solidification is that

$$|C_v(T_m - T_\infty)/L_v| \ll 1 \tag{6.3}$$

where C_v is the specific heat per unit volume of the melt, T_m and T_∞ are respectively the melting point (of a planar interface) and the temperature of the melt far away from the solid, and L_v is the latent heat per unit volume. In practice this condition is fulfilled unless $(T_m - T_\infty)$ is large (i.e. of the order of hundreds of degrees).

In evaluating the stability there are essentially two conditions which may be considered important. The first is the same as that used for the planar interface case in the previous chapter; that is, for stability we must have

$$\dot{\delta}/\delta \leqslant 0. \tag{6.4}$$

However, the real issue of concern is not whether the absolute magnitude of the perturbation grows with time, but rather whether the non-sphericity of the nucleus increases with time. This latter condition corresponds to the value of δ relative to R increasing with time. Thus of greater importance is the more easily satisfied condition for relative stability

$$\dot{\delta}/\delta \leqslant \dot{R}/R. \tag{6.5}$$

Note that it is quite possible for (6.5) to be satisfied even when (6.4) is not. Equation (6.4) is sometimes known as the absolute stability condition though this should not be confused with the rather different sense of the phrase as used in the last chapter. Equation (6.5) is of course, a necessary but not sufficient condition for (6.4).

Working through the analysis, it is found that, as would be expected from the results obtained in the previous chapter, there are two processes opposing one another in satisfying the stability criteria. On the one hand thermal diffusion tends to destabilise the interface by transferring heat concentration ahead of the protruberances towards depressions in the interface, whilst, on the other hand, capillarity opposes the growth of protruberances. This result may be expressed by writing down the critical radius for each harmonic (of order l) above which the sphere will be unstable to perturbations of that harmonic. Neglecting interface kinetics, the radius for absolute stability R_a is given by

$$R_a = R^*\{1 + \tfrac{1}{2}(l + 2)(1 + \kappa l)\} \tag{6.6}$$

where

$$\kappa = 1 + (K_S/K_L), \tag{6.7}$$

K_S and K_L being the thermal conductivities of the solid and liquid phases; and for relative stability the radius is

$$R_r = R^*\left\{1 + \frac{1}{2}\left[\frac{l-1}{l-2}\right](l + 2)(1 + \kappa l)\right\}. \tag{6.8}$$

R^*, which is a function of the overall undercooling of the solid relative to the melt far away from it and the capillarity constant, is simply the critical radius for nucleation at this undercooling; i.e. the radius above which the solid will grow in the situation considered in chapter 2.

It is immediately clear, as would be expected, that the critical radius for stability is smallest for the lowest harmonics (corresponding, in effect, to the longest wavelength perturbation and hence the weakest capillarity effect). However, the case of $l = 1$ need not be considered; this simply has the effect of translating the sphere and is not of interest. Indeed this is reflected in the factor $(l - 1)$ which appears in the product for $\dot{\delta}_l$ which is the rate of change of δ with time for a specific harmonic denoted by l. For $l = 2$ we see that a real value of R_a comes out of (6.6) but that R_r according to (6.8) is infinite. Thus, while the sphere is unstable in an absolute sense to perturbation tending to change the shape to an ellipsoid, it is always stable *relatively* to this perturbation at any finite size. The first l value of interest is therefore $l = 3$. Taking $(K_S/K_L) = 1$ as an example, this leads to values for R_a and R_r of $18.5R^*$ and $36R^*$ respectively. For higher l values $(R_r/R_a) \simeq (l - 1)/(l - 2)$ so that as l increases the difference between relative and absolute stability conditions rapidly diminishes. Evidently the largest and most important difference between the two criteria occurs in the $l = 2$ case.

In the last chapter it was seen that introducing interface kinetics effects slowed down the rate of growth of perturbations in the unstable region so that in some cases they may never be observed in a finite time. Clearly, in the case of the spherical morphology where we are concerned with relative stability this effect may be far more important; a slowing down of the growth rate of a particular harmonic of a perturbation may well mean that relative stability appears if a suitable interface kinetics term is included where it did not exist without it. Coriell & Parker (1967) have studied this situation. Their calculations are simply an extension of the stability theory for the growing sphere including the effect of growth kinetics. This is done explicitly by assuming that the kinetics follow one of the two growth laws generally thought to be applicable to solidification; that is either that the growth rate is directly proportional to the local interface supercooling (i.e. the difference between the actual interface temperature and the equilibrium temperature of an interface of the same shape), or that it is proportional to the square of this super-cooling. These two growth laws are generally associated with metallic (non-singular) and faceting growth respectively (see chapter 8). Taking the constant of proportionality as μ in each case we can define a quantity α for each law as

$$\alpha = (K_L/L_v\mu R^*). \tag{6.9}$$

Coriell and Parker then tabulated the instability points for various values of α in each case. The results are shown in tables 6.2 and 6.3.

Table 6.2

α_1	$R_r(3)/R^*$	Minimum $R_r(l)/R^*$	l at minimum
0	3.60 (10^1)	3.60 (10^1)	3
10^{-3}	3.60 (10^1)	3.60 (10^1)	3
10^{-2}	3.61 (10^1)	3.61 (10^1)	3
10^{-1}	3.68 (10^1)	3.68 (10^1)	3
10^0	4.36 (10^1)	4.36 (10^1)	3
10^1	1.09 (10^2)	9.05 (10^1)	4
10^2	7.40 (10^2)	4.01 (10^2)	7
10^3	7.04 (10^3)	2.76 (10^3)	13
10^4	7.00 (10^4)	2.32 (10^4)	27
10^5	7.00 (10^5)	2.14 (10^5)	$\sim$ 56
10^6	7.00 (10^6)	$\sim$2.07 (10^6)	> 100

Table 6.3

α_2	$R_r(3)/R^*$	Minimum $R_r(l)/R^*$	l at minimum
0	3.60 (10^1)	3.60 (10^1)	3
10^{-3}	3.69 (10^1)	3.69 (10^1)	3
10^{-2}	3.87 (10^1)	3.87 (10^1)	3
10^{-1}	4.52 (10^1)	4.52 (10^1)	3
10^0	7.16 (10^1)	6.62 (10^1)	4
10^1	2.30 (10^2)	1.48 (10^2)	5
10^2	1.65 (10^3)	5.62 (10^2)	7
10^3	1.58 (10^4)	3.31 (10^3)	13
10^4	1.58 (10^5)	2.54 (10^4)	25
10^5	1.58 (10^6)	2.24 (10^5)	52
10^6	1.58 (10^7)	$\sim$2.11 (10^6)	> 100

As expected in each case, as α increases (and hence the degree of kinetic control increases) the radius at which relative instability occurs increases. More particularly, however, recourse to the rather limited amount of experimental data shows that the quantity μ is generally very much larger for linear growth law materials than for square law materials. Typically, they differ by factors of the order of 10^5. Thus under similar undercooling conditions metals should have a much smaller critical radius than non-metals. As an example Coriell and Parker calculate the critical radii for salol and tin (as typical materials of the two types) assuming the same undercooling of 5 deg. For salol this gives a radius of 0.5 cm which is consistent with the stable faceted crystals of this kind of size which are commonly observed, whereas for tin the calculated radius is 3 μm. In view of this latter result it is not surprising that dendritic growth is always observed when a melt of tin is undercooled. Other experimental data derived from measurements of the radii of dendrite growth tips implies that this size is probably of the right order. This point will be returned to later in the chapter.

These analyses therefore present the picture of a nucleus, initially assumed to be spherical, gradually becoming distorted as it grows. In the case of faceting materials the growth kinetics appear to stabilise the shape up to sizes of the order of centimetres, but in the case of metals the deformation becomes serious at sizes only a few times the critical radius for growth under the ambient supercooling. So far no mention of crystalline anisotropy has been introduced into the analysis. Clearly, the effects of anisotropic interfacial free energy and interface kinetics will tend to stabilise even further these harmonics which are not represented in the crystalline symmetry. Thus while on the isotropic model the $l = 3$ harmonic is always the first to be unstable if kinetic control is weak it may well be that for certain symmetries it is a higher harmonic which will be the first to be unstable in a real experiment. For instance the observed orthogonality of dendrite directions observed in cubic materials implies that their four-fold symmetries are allowing only harmonics which are multiples of four to grow. Unfortunately no detailed calculations of this type have been performed because these could not be pursued to a conclusion. There is an absence of experimental values for the anisotropy of interfacial free energies and interface kinetics measurements of anisotropy are sadly lacking; a similar sparcity of kinetics results will be revealed in chapter 8. However, Cahn (1967) has shown by general arguments that anisotropy of the interfacial free energy in particular will cause the distorted nucleus to reflect this when it goes unstable, which is in agreement with our intuitive arguments. In particular, he concludes that one of the main effects of anisotropy will

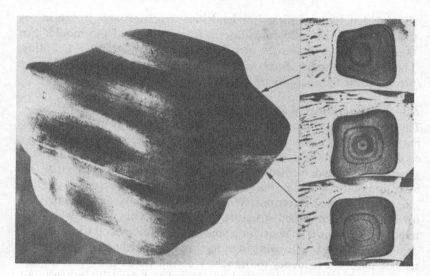

Fig. 6.1. Composite model and sections of 'predendrite' in Al–0.1% Cu alloy after Kiss and Biloni (see ref. 31 of Biloni (1968)). Micrographs × 55.

be to impose on the shape initial perturbations reflecting the anisotropy, which will then be enhanced by the diffusion fields.

In addition to the general support given to the theory by the experimental observation that metals usually grow dendritically and nonmetals are relatively more stable, there are some experiments on the structure in castings of pure metals by Biloni (1968) which also support the general picture of the way in which instability occurs. Usually in the freezing of metals nucleation occurs at the surfaces of the mould and dendritic growth proceeds into the supercooled melt while heat is extracted through the solid to the walls. In alloy castings constitutional supercooling effects further ensure that the dendritic morphology is observed. It appears possible, however, that some nucleation can also occur within the melt (presumably heterogeneously on small impurity particles, assuming that the supercooling is much too small for homogeneous nucleation). In a pure metal system these nuclei can be expected to grow quite large (i.e. to a microscopic size rather than to atomic dimensions) before instability would be predicted, because of the small undercooling present around them, and the critical radius is much larger than the few atomic radii associated with homogeneous nucleation. However, slight additions of impurity (0.1 per cent) produce constitutional supercooling effects as the nuclei grow and thus encourage instability at smaller radii. These effects were observed by Biloni who showed that some of these 'predendritic' structures are 'frozen in' by the growth of the solid spreading in from surface of the mould. For example, fig. 6.1 shows the structure observed in Al–0.1 ± Cu alloy. There is a striking resemblance of this structure to the kind of shape predicted for the early stages of dendritic growth.

6.3   The stability of a growing rod

The stability theory applied to a spherical morphology explains why a nucleus does not grow into a large spherical mass of solid when super-cooling is present. In particular we can see that the spherical shape will generally become deformed quite early, and it seems reasonable that the extension of this would be the development of spikes in certain crystallo-graphically defined directions. We do not as yet have any explanation for the detailed branched structure of dendrites. One way in which we might achieve this is to consider a spiked morphology and to think of a dendrite arm as a rod growing in radius, and to see what stability theory tells us about the tendency to branch. We shall see a little later that this may not in fact be a very good physical model of dendritic growth and therefore may not be so useful in the understanding of dendrite mor-phology. Analyses of rod stability do, however, provide further insight into stability effects, and, more important, lead to one of the most significant experimental tests of interface stability theory.

By now it should be clear how to approach the problem of considering the stability of a growing rod. Generalised perturbations can be analysed in terms of sinusoidal perturbations along the length of the rod and cylindrical harmonics round the circumference. As might be expected the basic results are similar to those for a sphere; instability in the cylindrical harmonics sets in when the radius reaches a critical value, and relative stability as well as absolute stability is a relevant criterion. However, whereas in the case of spherical morphology, the ratio of critical radius for instability to critical radius for homogeneous nuclea-tion was a number, dependent only on the harmonic of distortion being considered, in the cylindrical case this ratio is also a function of the driving force for growth. This is a result of the boundary conditions which must be chosen to allow Laplace's equation to be a useful approximation to the time-dependent growth. One particularly interest-ing effect which should be noted is that in the absence of a diffusion field a rod is an inherently unstable morphology in spite of (indeed, *because of*) the interface tension effect. This is because the surface-to-volume ratio of a rod is decreased if perturbations of wavelength greater than the cross-sectional circumference grow. A simple proof of this result is given by Pfann & Hagelbarger (1956).

Several analyses of the growing rod problem have been carried out to include not only considerations of kinetic effects and surface tension, but also a small degree of anisotropy in these. For example, Kotler & Tiller (1967) have tried to determine the fastest growing perturbation along the axis of the rod. Writing the perturbed interface as

$$r = R + \delta \sin (k\phi + (k_z z/R))$$

where z is measured along the rod and ϕ is an azimuthal angle, they were concerned with finding the fastest growing value of the wavelength λ_z associated with k_z. They argue that this provides a useful model for dendrite growth and thus that these values of λ_z should correspond to

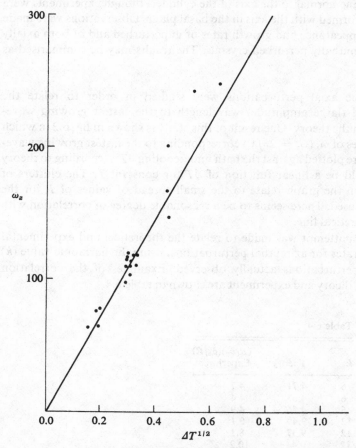

Fig. 6.2. Relationship of wavelength (via ω_z) and bath undercooling for z-perturbations in the experiments of Coriell & Hardy (1969).

the branch spacing of dendrites grown at a similar bath undercooling to that used in the model calculations. These calculations are not without success and will be referred to again later in the chapter.

The other main contribution to the problem of the stability of growing rods has been that of Coriell and Hardy who have supplemented their theoretical calculations with some very elegant experiments. As these are probably the most detailed attempts to test morphological stability theory, it is worth describing their results in some detail. The experiments consisted of growing cylindrical single crystals of ice initially about 1 mm radius in supercooled water baths under very carefully controlled conditions. The crystals were usually prepared with their basal plane normal to the axis of the cylinder, though experiments were also performed with the axis in the basal plane. Observations were made on the appearance and growth rates of unperturbed and of both axially and azimuthally perturbed crystals. The results may be summarised as follows:

(1) The axial perturbations were studied in order to relate the observed (large amplitude) wavelength to the fastest growing wavelength in the theory. One result of this study is shown in fig. 6.2 in which the values of ω_z ($\omega_z = 2\pi/\lambda_z$) corresponding to the fastest growing wavelength are plotted against the bath undercooling ΔT. According to theory ω_z^2 should be a linear function of ΔT for constant T. The clusters of points on the graph relate to the small spread of values of R for the crystals used. There seems to be a reasonable degree of correlation with the theoretical line.

(2) An attempt was made to relate the theoretical and experimental growth rates for azimuthal perturbations, using the harmonic value (k) of the perturbations actually observed. Examples of the correlation between theory and experiment are shown in table 6.4.

Table 6.4

k	Theory	$(R/\delta_\phi)(\dot{\delta}_\phi/\dot{R})$ Experiment
6	4.71	4.7
6	4.63	4.4
6	4.56	4.9
12	9.49	9.1
12	9.37	8.1
12	9.44	10.2
12	9.37	9.2

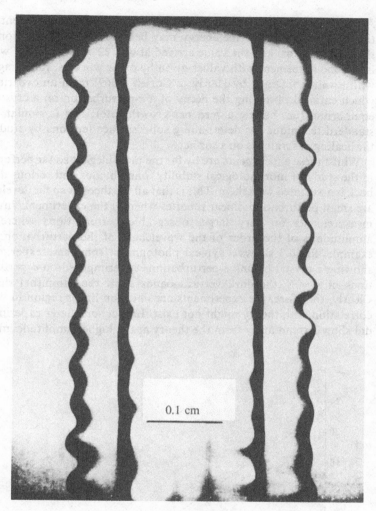

0.1 cm

Fig. 6.3. Example of ice cylinder with z- and ϕ-perturbations ($k = 6$) superimposed, after Hardy & Coriell (1968).

(3) A similar attempt was made to relate the observed growth rate for z-perturbations with theory. As ϕ-perturbations occur before the z ones are observed, the theoretical calculations also allow for ϕ-perturbations of the observed harmonics. A similarly satisfactory correlation was observed in these experiments.

In order to correlate these experiments with the theory it is necessary to know a value for the interfacial free energy of the ice–water interface.

Alternatively, by anticipating good correlation between experiment and theory, a value of this free energy may be deduced. This was done in these experiments and the value arrived at was 22×10^{-3} J m^{-2} which is in good agreement with values given by other workers. It also agrees with a value obtained by Hardy & Coriell (1969) from an experiment which entailed observing the decay of a z-perturbation on a crystal in equilibrium (i.e. having a zero net growth rate); this is similar to a standard technique for determining solid surface tensions by studying the healing of scratches on a surface.

Whilst these experiments are by far the most elegant so far performed in the study of morphological stability, one obvious and serious drawback is associated with them. This is that all the theories so far developed are small perturbation, linear theories whereas the experiments involve measurements on very large, observable perturbations where the amplitude is of the order of the wavelength of the perturbation. For example, fig. 6.3 shows a typical photograph from these experiments showing a crystal having z-perturbations superimposed on ϕ-perturbations of $k = 6$ (the dark vertical bands show the azimuthal shape). Clearly, therefore, the experiments are in a non-linear region in which correlation with theory might not exist. In fact some later experiments did show a trend away from the theory at the higher amplitude growth

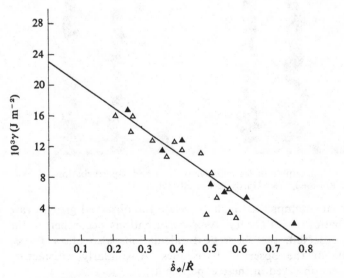

Fig. 6.4. Predicted value of the ice–water interfacial free energy versus perturbation growth rate in the experiments of Hardy and Coriell (1969). ▲ $k = 6$, △ $k = 12$.

rates. Uncertainty in the value of the parameter γ, the interfacial free energy, allows some degree of freedom in comparing theory with experiment. However, if the results are extrapolated towards zero growth rate of perturbation, the interfacial free energy arrived at is the one quoted above which gives the best agreement with other results. This can be seen most clearly in fig. 6.4 which shows the variation in value of γ as implied by the correlation of theory with experiment as a function of relative perturbation growth rate $(\dot{\delta}_\phi/\dot{R})$. Thus the most satisfactory correlation is observed in the limit at which the theory should be applicable.

6.4   A steady-state description of a dendrite cap

Before any quantitative description of stability effects had ever been attempted many workers were interested in the problem of describing the shape of the dendrite in the region of its tip. As dendritic growth is associated with large supercooling and high growth rates it has always been assumed that the shape of the cap of a growing dendrite should be such as to allow the most rapid dissipation of the heat fields generated around it. Such a model initially neglects the branched structure of dendritic growth but rather concentrates on the shape of it; moreover, as branching is essentially not included in the analysis it is natural that the shape of the cap should be considered to be constant in time in a frame of reference moving with the constant velocity of the cap.

The first serious attempt to describe the tip shape was made by Ivantsov in 1947 assuming it to be a parapoloid of revolution. This had been suggested previously (Papapetrou, 1935), and indeed a glance at a photograph of a growing dendrite confirms that this is a reasonable first approximation. Ivantsov showed that this shape was consistent with a model of dendritic growth in which the interface was assumed to be isothermal. He then investigated the relationship between the growth velocity V, the radius of curvature of the tip ρ and the bath undercooling ΔT. He found that for each value of ΔT the product $V\rho$ was defined as some constant, but that he could not separate these variables. This would imply that there is no unique value of V and of ρ for some value of ΔT; moreover, it implies that V could in effect be made infinite simply by letting $\rho \to 0$. Clearly this is not a physical possibility; one would naturally expect that both V and ρ are uniquely defined by the external conditions.

The reason for the failure of Ivantsov's description is simply that he did not introduce sufficient physical conditions into the problem. In particular both interfacial tension and interfacial kinetics play an important role; interface kinetics obviously limit the rate of growth of the material and both these effects introduce temperature variations into the interface. Firstly there is a variation in equilibrium temperature as the radius of curvature changes due to capillarity, and secondly if the shape is to be a steady-state one, the rate of growth and hence associated kinetic undercooling of the interface must also be a function of position. Thus the isothermal dendrite cannot be a realistic model. The problem now is how to describe the shape. Temkin (1960) dealt with this by continuing to assume that the shape is a paraboloid of revolution, but taking into account the non-isothermal nature of the interface resulting from these two effects. The justification for keeping the same shape is simply that it is likely to be a convenient approximation and that for a sufficiently small part of the cap a paraboloid must necessarily become a good description; just how large this small part should be remains an unanswered question. Pursuing the analysis, Temkin found that for a given undercooling, there is now a maximum in the graph of V against ρ. The choice of the probable unique values of V and ρ is now easier as the maximum is the only unique point in the relationship. The idea that a dendrite should adopt a form giving maximum growth rate is by no means a new idea although it has been the subject of considerable argument. As Temkin points out this seems likely to be the stable solution in that if any perturbation should grow ahead of the tip it must be 'swallowed up' by the growing dendrite which is already growing at the maximum growth rate permitted; any smaller, or larger value of ρ corresponds to a lower growth rate. It is interesting to note in passing that the maximum growth rate predicted by Temkin's analysis is always appreciably smaller than that given by Ivantsov for the same value of ρ. In order to complete the analysis Temkin makes the simplification that $V\rho/a_j \ll 1$ where a_j is the thermal diffusivity of either phase.

Despite the several approximations made, Temkin's analysis is quite successful. In particular, Kotler & Tiller (1968) in a recent review used the analysis to derive the interfacial free energy and interface kinetics term for ice and tin from published data on experiments on the free dendritic growth of these materials from pure melts. The kinetic co-efficients cannot readily be compared with other data and so the values obtained can only be said to be reasonable. However, the free energy values are in quite good agreement with other measured values. In particular, the value for ice was $(20 \pm 2) \times 10^{-3}$ J m^{-2} which is in good agreement with the value given by Hardy and Coriell mentioned in the previous section, and indeed with most other values apart from that obtained by Turnbull from the results of nucleation experiments $(32 \times 10^{-3}$ J m$^{-2})$. In view of the marked anisotropy of growth and

our earlier comments about the form of average result given by nucleation data this discrepancy is not altogether surprising.

6.5   Dendritic branching

So far we have shown that in the presence of supercooling a small
nucleus will develop spikes, and that we expect these spikes to take on
the general form of paraboloids of revolution. The study of the stability
of a rod growing radially in a diffusion field has shown that this morphology is also unstable, so that one might construct a model for the
branching behaviour of dendrites by considering that instabilities occur
on the sides of the spikes as they grow outwards into the melt. It is
interesting to see if the analysis can be taken a little further so that we
can actually see how these branches appear and grow. In particular we
might ask whether it is possible to predict the spacing of these arms as
observed during, or after, growth.

The first model, and one that we have already discussed, is simply to
think of a dendrite as a growing cylinder. This is a situation which can
be analysed fairly straightforwardly, and attempts have been made in
this way to predict arm spacings. However, the problem here is to decide
how realistic it is to use this approximation of a cylindrical shape for a
spike which in its unperturbed form we believe to be roughly para-
boloidal. If the instability appears first a long way behind the penetrating tip of the dendrite this might be a good approximation. On the
other hand if the instability originates near the tip then the main shape
transformation towards the branched structure may occur before a
cylindrical model could ever be relevant. The very limited amount of
experimental data involving direct observation of dendrite growth as it
occurs indicates that perturbations are visible at a distance of a few tip
radii behind the tip (see for example fig. 6.5). This is rather near and
would at least lead us to feel that a fuller analysis would be more
satisfactory.

In fact Kotler & Tiller (1968) have tackled the formidable problem of
investigating the stability of the 'dendrite' as described by Temkin's
equations. Unfortunately, as pointed out by Sekerka (1968) there is an
inherent danger in this analysis. It has already been pointed out that a
paraboloid of revolution is not (and cannot be) the exact correct
solution to the unperturbed heat flow problem with a non-isothermal
interface. In analysing the stability they are therefore starting off with a
configuration which *cannot be* stable because it does not satisfy the
unperturbed steady-state conditions. Incidentally it is also worth com-
menting that even if the analysis were valid, there is considerable

Fig. 6.5. Growing dendrites of pure tin seen on a surface of the solid–melt system, × 22. After Glicksman & Schaffer (1966).

difficulty in describing the perturbed interface. Fig. 6.6 shows the two possible forms of perturbation which might be considered and which are termed by Kotler & Tiller the 'standing wave' and 'travelling wave' models. A little thought about the problem will show that while the 'standing wave' solution is easier to describe, being stationary in the frame of the steady-state growing dendrite tip, it is an unlikely form of development since it would involve certain parts of the dendrite solidifying and then melting back, perhaps several times. The 'travelling wave' solution allows for a steady development of the side-arm perturbations but it is not an allowed form of solution in the analysis of Kotler & Tiller. Thus it is only too clear that a proper solution to the problem is not readily forthcoming.

One final point is worthy of consideration. Using a simple-minded approach to the problem we can visualise a fine spike appearing on an unstable nucleus above the critical size. Now Temkin's analysis tells us that there is an optimum tip curvature which allows the fastest growth rate. Thus if this spike has a smaller tip radius than the optimum, it will slowly enlarge and grow faster. Now suppose that we can consider the enlarging tip like a growing sphere and that the critical size for instability is a sphere smaller than the optimum tip radius (a supposition which, while not being unreasonable, is by no means obvious *a priori*).

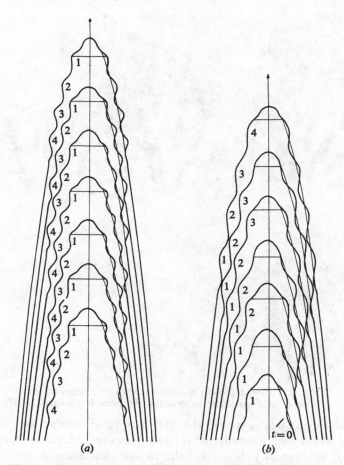

Fig. 6.6. The two types of dendrite perturbation discussed by Kotler & Tiller (1968): (a) 'standing wave', (b) 'travelling wave'. The development of the dendrite during growth is shown as time, t, increases.

This being the case, when the tip radius reaches the critical size it will become unstable and a new smaller spike will form on the tip and pass through a similar development cycle. We therefore have a mechanism for dendrite growth and branching; each time a new spike grows out the old tip can develop sideways into branch arms. The interesting feature of this mechanism is that it is not a steady-state form as has been previously supposed; it is, in fact, the 'travelling wave' sort of solution, though somewhat exaggerated. It now seems that this jerky form of dendritic growth does actually occur. In fact it was predicted by a computer simulation of dendritic growth by Oldfield (1968), although

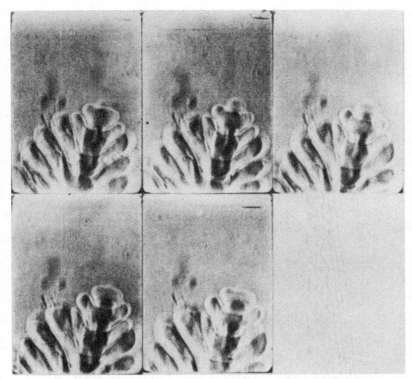

Fig. 6.7. Pulsed growth mechanism seen in a growing dendrite of an organic metal analogue material (after Morris & Winegard, 1967), × 305.

at the time it was suggested that this may be an artefact of the finite steps which such a simulation follows. However, Morris & Winegard (1967) have recently shown that this form of branching certainly exists and fig. 6.7 shows a cycle in this 'pulsed development'. On the other hand observations under different conditions do not always appear to show this effect, so it cannot yet be said that the branching of dendrites is completely understood.

6.6   Cast structures and dendrites

Undoubtedly the main practical reason for wanting to understand dendritic growth lies in the need to control the structure of cast materials. Alloys almost invariably solidify dendritically (for reasons which, in general terms, should now be clear) and so, for instance, the degree of local segregation in a casting is controlled by the spacing of dendrite arms in the final casting. The full description of these effects is outside the scope of this book; in any case it is dealt with in Chalmer's (1964)

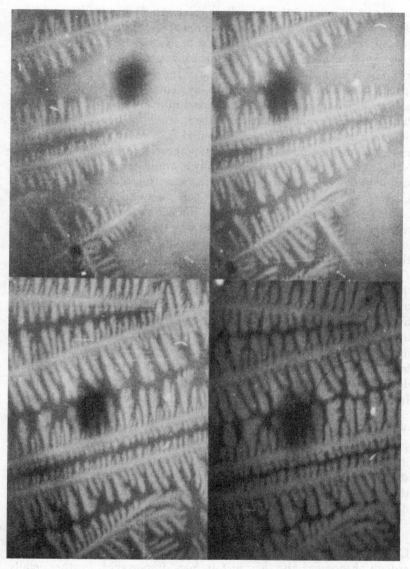

Fig. 6.8. Consecutive photographs of the same area of a specimen showing dendritic growth of a sodium–potassium alloy by transmission ultra-violet microscopy, ×25.

book and in many other metallurgical texts. However, it would be rather satisfying, having reached a degree of understanding of dendritic growth, to feel that all these problems in the metallurgical world were now understood. To show that the problems are even greater than they might have seemed so far, let us consider the problem of predicting the spacing of dendrite arms in a cast alloy. This is a problem which has been tackled empirically many times and in many systems. Parameters deemed to be relevant and related to the spacings include rate of cooling, time to final solidification and the initial 'superheat' (i.e. temperature above the melting point – not true superheating) of the poured alloy. Now to solve this problem theoretically it would appear to be necessary simply to determine the fastest growing wavelength of perturbation on a growing cylinder of the material; this ought to give a reasonable answer, on the basis of experience in simpler systems. A glance at the series of photographs in fig. 6.8 shows why this approach is doomed to failure. The spacing of side branches near the tip of the dendrite is much closer than it is at the same position when the tip has passed further on. This is because as the growth proceeded outwards from the main arm, the rate of growth slowed down, some side arms grew at the expense of others, and so the whole structure coarsened. In many cases the original fine branch spacings can no longer be detected because these small branches have been overgrown by the main arm. This shows that the final structure of a casting is not simply determined by the original branch spacings generated from the perturbed rod structure, but also depends very considerably on the subsequent conditions of growth. Indeed, a structure left at a fixed temperature, with no net growth, will coarsen markedly due simply to the driving force of surface free energy minimisation. In fact recent experimental work suggests that the total solidification time is probably a more important parameter than the initial growth rate. Evidently a proper theoretical description of this situation from first principles would involve a full non-linear analysis of the development of alloy dendrites, including the effects of neighbouring dendrites, and would be a formidable task.

Further reading
Chalmer's book (1964) is useful as a general background to dendritic growth phenomena and their metallurgical importance. Further details of the other points dealt with in the chapter are only readily found in the original papers. In particular, Hardy & Coriell (1969) review their own rather beautiful experiments on morphological stability in the growth of ice cylinders, and the paper by Kotler & Tiller (1969) usefully reviews the heat-flow theories of dendrite cap shapes.

7　Eutectic Growth

7.1　Introduction

Fig. 7.1 shows schematically a typical simple eutectic phase diagram. At concentrations near the pure A or B ends of the diagram, the situation is the normal one for a distribution coefficient $k < 1$. On the left-hand side a solid (α-phase) which is predominately A with a small concentration of B in solid solution, is in equilibrium with liquid richer in B at some temperature below the melting point of pure A. On the right-hand side a similar situation exists with the solid (β-phase) being rich in B. However, the liquidus line is not a smooth gradation between the two ends but has a singularity at the so-called eutectic point (eutectic is from the Greek meaning most fusible – i.e. it has the lowest possible melting point of the system). If a liquid of eutectic composition is solidified then the phase diagram shows that rather than a single solid phase resulting, both the α- and β-phases should be produced together in equilibrium and no net change in composition in the liquid should occur (as it does in single phase solidification due to the segregation at the interface). This is therefore a basically different effect from that of single phase solidification which is dealt with in the rest of this book, and for this reason alone is well worth special attention.

There are good reasons for the recent upsurge of interest in eutectics. This is the result of the major development of composite materials in the past few years. The basic idea, which is not in itself new, is to try to produce a material which consists of very strong but possibly brittle fibres embedded in a ductile matrix. Depending on the properties of the interphase boundary it is possible to produce a material which has much of the strength of the fibres but without the brittle properties generally associated with the fibres by themselves. The best-known example of such a material is fibreglass reinforced plastics whose strength and durability are well known. In extending this idea to develop even stronger materials (for example, the recent development of carbon fibres) there is often a problem in the handling of very fine fibres and embodying them in a suitable bending matrix, particularly as the fibres must often be aligned to give the desired mechanical properties. This is where eutectic solidification may supply a new approach, particularly in the case of intermetallic compound fibres. If a eutectic liquid is solidified unidirectionally, then the two phases that separate out may well grow

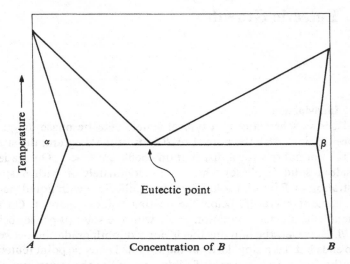

Fig. 7.1. A simple binary eutectic phase diagram.

as aligned rods or lamellae normal to the planar isotherms of the system. Thus, in principle at least, a method exists for manufacturing the complete composite material in a single stage process.

There is therefore good reason why we should single out binary eutectic solidification from the general area of polyphase solidification to see if the basic factors controlling the processes of growth can be understood. However, even in the restricted field of binary eutectics, the variety of structures encountered is vast and bewildering. Many attempts have been made to classify these structures but most of them are empirical and have in many cases been shown to be spurious. One classification which has recently gained ground in one form or another and which has the merit of being based on a physical model of the growth mechanisms is that of Hunt & Jackson (1966). They suggest that the structures observed in eutectics can be related to the faceting or non-faceting behaviour of the constituent phases. Thus there are three basic categories: both phases faceted, one faceted and one non-faceted, and both phases non-faceted. Discussion in this chapter will be restricted mainly to the non-faceted/non-faceted eutectics whose structures are most easily understood. A single faceting phase greatly complicates the structure and the faceted/faceted eutectics have a highly irregular structure; indeed, in some of these materials at least, it seems that the two phases grow largely independently of one another but at the same time, and are therefore much less interesting than the other systems which exhibit true cooperative growth of the two solid phases. In using this classification of eutectics it is worth remembering that the tendency to form a faceted or non-faceted interface in a pure solid grown from its own melt is not necessarily followed when a solid solution grows from an alloy. An example of this was seen in the silver-rich phase of a silver–bismuth alloy in chapter 3. For this reason the factor determining the structure is probably the faceting or non-faceting behaviour of the eutectic solid phases grown from the eutectic liquid rather than the pure solid α-factor.

7.2   Lamellar eutectics

Restricting the discussion now to non-faceted/non-faceted eutectics, one of the most common structures encountered is the lamellar one, an example of which is shown in fig. 7.2. The material consists of regular parallel sheets of one solid phase interleaved with sheets of the other phase. If such a eutectic is grown uniaxially under stable conditions from pure constituents then these lamellae can persist in a regular fashion over several centimetres of the ingot. It was at one time thought that the lamellae lay parallel to the isotherms (i.e. normal to the growth direction) and that growth proceeded by layers of α- and β-phase forming alternately on the solid–liquid interface as the liquid became enriched first in A and then in B. However, it is now known that the lamellae lie parallel to the growth direction. Thus an array of lamellae of α- and β-phases grow simultaneously edgewise into the liquid. Presumably the original nucleation of solid occurs either by the creation of a duplex nucleus consisting of α- and β-phase, or one phase nucleates first and then, as this grows, the local enrichment of the major constituent of the other phase in the surrounding liquid causes this second phase to nucleate, presumably heterogeneously, on the solid–liquid interface of the first phase. Of these possibilities the latter is obviously intuitively more reasonable. In either case we would expect that if a configuration with low interfacial free energy exists by suitable matching of similarly spaced planes in the two adjacent solid phases, then this relative orientation would be produced between the phases and a well-defined orientation relationship would exist between them. It was thought that new lamellae could then nucleate successively on the alternate phases until a suitable array had built up to allow further growth to continue by the normal edgewise mechanism just described. However, there are difficulties about such a process of successive nucleation and, in particular, it is difficult to see how highly regular structures of the kind observed would result. Moreover, some fairly random misorientations of the phases would be expected (having regard to earlier comments in chapter 2 about

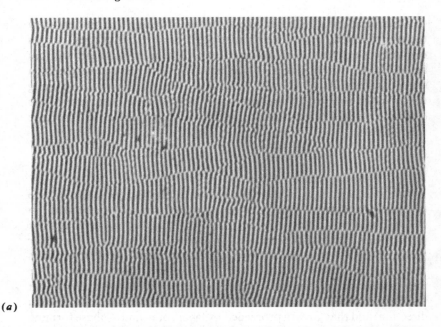

(a)

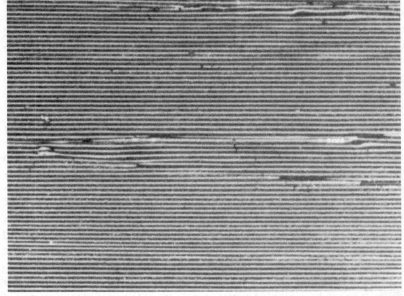

(b)

Fig. 7.2. (*a*) and (*b*): A regular lamellar eutectic shown in transverse and longitudinal sections (LiF–CaF$_2$). (*a*) ×190, (*b*) ×210. (Courtesy Dr M. W. A. Bright.)

the variety of orientations observed in experiments on heterogeneous nucleation). Experiments show, however, that as well as being regularly spaced, all the lamellae of each phase throughout a 'eutectic crystal' are almost identically oriented crystallographically. It therefore seems fairly clear that there must be some 'bridging' mechanism whereby new lamellae can be formed from old ones so that the whole of each phase is interconnected. Perhaps rather surprisingly in the light of this, it has recently been shown that there are, in fact, small but well-defined misorientations between adjacent lamellae of the same phase (Weatherly, 1968; Davies & Hellawell, 1969).

Models of the shape of the solid–liquid interface during growth of a lamellar eutectic assume that the interface is isothermal; considerations of heat flow across the interface show that this must generally be true to better than 0.01 deg (Jackson & Hunt, 1966). If we consider two adjacent lamellae of α- and β-phases growing into the melt, then it is clear that the α-phase must reject excess of component B and the β-phase must reject excess of A into the melt. For growth to proceed these components must diffuse laterally across the interface and so a concentration profile like that shown in fig. 7.3(a) must exist in the liquid in front of the lamellae. Evidently this variation in composition across the boundary produces a variation in local melting temperature across the interface, this effect must be offset by the curvature of the boundary. This may be expressed in terms of an 'undercooling' of the interface (i.e. the difference between the actual temperature and the equilibrium eutectic temperature), ΔT, where

$$\Delta T = \Delta T_D + \Delta T_C + \Delta T_K = \text{constant across the interface.} \quad (7.1)$$

Here ΔT_D is the 'undercooling' due to the variation in composition from the eutectic composition, ΔT_C is the 'undercooling' due to capillarity, and ΔT_K is the 'undercooling' due to interface kinetics. (Notice that the term undercooling is used rather loosely in discussions of this kind. It is often used to mean the difference between interface and bath temperature but it can also be used to mean specifically the kinetic driving force (i.e. ΔT_K only). The use of this term must therefore always be approached with caution.) For the case of non-faceted (metal-like) growth it is usually safe to neglect ΔT_K altogether as being much smaller than the other contributions in all practical situations of interest. Thus the diffusion of the rejected solute determines the interface curvature and hence the interface shape.

However, several other factors needed to completely define the interface shape have still to be determined. In particular, the actual spacing of pairs of lamellae has not yet been specified. Experiment shows that

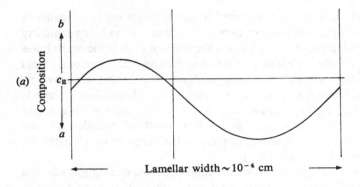

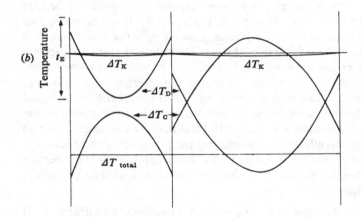

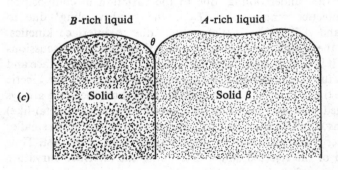

Fig. 7.3. (*a*) Schematic concentration profile of *B* ahead of the interface in the liquid. (*b*) Contributions, ΔT_D, ΔT_C and ΔT_K to the total interface undercooling. (*c*) Predicted interface shape (after Jackson & Hunt, 1966).

the spacing is a well-defined function of the growth rate, and that as the growth rate increases the spacing decreases. This is what might be expected intuitively: in the absence of other considerations there would be a tendency to develop very large lamellar spacings because this minimises the total free energy of the system by reducing the area of α–β interface; on the other hand large spacings are restricted by the necessity for diffusion of excess of components A and B to occur across the interface between the α- and β-lamellae. Thus, as the growth rate increases the diffusion required to maintain the lamellar structure can only be achieved by a reduction in lamellar spacing.

A formal solution to the problem has been given by Jackson & Hunt (1966). They extended the ideas put forward by previous authors (Zener, 1946; Tiller, 1958) and solved the diffusion equations in conjunction with the isothermal interface condition. They simplified the problem by first solving the diffusion equations for a planar interface. Introducing parameters for the α-phase–liquid, β-phase–liquid and α–β interfacial free energies, they found that an approximate solution is given by

$$\frac{\Delta T}{m} = v\lambda Q^{\mathrm{L}} + \frac{a^{\mathrm{L}}}{\lambda} \qquad (7.2)$$

where v is the growth rate, λ the interlamellar spacing (i.e. the width of an $\alpha + \beta$ lamellae pair) and m, Q^{L} and a^{L} are constants. In particular m depends on the liquidus gradients for each phase, Q^{L} is a function of the relative sizes of α and β lamellae and the volume fraction ratio together with the diffusion coefficient, and a^{L} is determined by the volume fraction, liquidus gradients and the various interfacial energies in the system. Fig. 7.4 shows (7.2) plotted out for some value of v. Clearly the values of λ and ΔT are not yet fixed uniquely by v; the problem resembles that of the dendrite tip discussed in chapter 6 and some further condition is required to specify the problem. The most obvious solution is analogous to the one adopted (though not always accepted) in the solution of the dendrite growth problem, of taking the only singular point on the curve; i.e. that corresponding to a minimum of ΔT. In fact, like the dendrite growth solution, this is also the maximum growth rate solution in that the minimum in fig. 7.4 corresponds to a maximum in the graph of v against λ for constant ΔT. As before, this solution is rather unsatisfactory in its *ad hoc* nature but it does have the merit of producing the right sort of result. If the state fixed by the minimum is taken to be the correct solution then it is found that

$$\lambda^2 v = a^{\mathrm{L}}/Q^{\mathrm{L}}; \qquad \Delta T^2/v = 4m^2 a^{\mathrm{L}} Q^{\mathrm{L}}. \qquad (7.3)$$

These predictions that v is proportional to $(1/\lambda^2)$ and to $(\Delta T)^2$ have been verified by experiment.

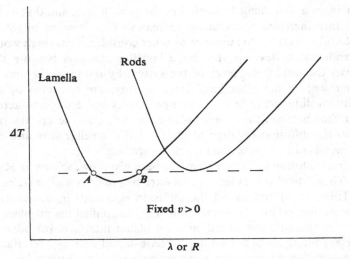

Fig. 7.4. Equation (7.2) for lamellae and a similar curve for rods plotted at some fixed growth rate.

In selecting the minimum as the relevant solution for the eutectic solidification problem, a difficulty arises which does not occur in the dendrite problem. A growing dendrite can easily modify the curvature of its tip and so can easily adopt the shape corresponding to the maximum growth rate. Changing the lamellar spacing in the eutectic is not so easy as it necessitates changes across the whole interface. For this reason an alternative approach to the selection of the actual growth conditions was proposed initially by Jackson & Chalmers (unpublished) and then developed formally by Jackson & Hunt (1966). The new approach is based on an experimental observation of the way in which lamellar eutectics can change their lamellar spacing by small amounts relatively easily. Fig. 7.5 shows the vehicle of this mechanism which is a lamellar termination, seen here as a schematic transverse section of a lamellar eutectic (i.e. a section normal to the growth direction). In this arrangement the mean lamellar spacing is less on the right of the termination than on the left; thus if the termination moves to the right the average spacing increases and if it moves to the left there is a decrease in average spacing. Now as the solution we are looking for should be a stable one this must correspond to a situation in which the position of the lamellar termination is also stable; i.e. in which there is no tendency for the termination to move either to the right or the left. The exact method of applying this idea used by Jackson and Hunt was as follows: they suggested that the degree of undercooling at which

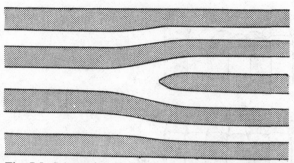

Fig. 7.5. Schematic lamellar termination as section normal to growth direction.

lamellae must grow should be the undercooling at which the termination is stable, and furthermore that this should correspond to the stable undercooling for the growth of cylindrical rods rather than lamellae. They arrived at a similar expression to (7.2) for a growing rod structure (see fig. 7.4 but note that the two curves have different x-axes, R being the rod spacing) and argued that the termination could adopt any shape and would therefore adopt a radius of curvature corresponding to the rod spacing which allows rod growth at the lowest possible under-cooling. Thus growth of lamellae should occur at the same under-cooling given by the position of the minimum in the curve for rods, and so at either point A or point B on the lamellar curve. The argument behind this idea is that, while it is difficult for changes to occur in the lamellar spacing, it is easy for the termination to adopt an optimum shape. As the termination will have diffusion fields surrounding it resembling those around half a rod, the minimum in the rod curve is a reasonable choice of condition. Now it remains to decide which of the solutions A and B is relevant, and this is done by considering further stability criteria following arguments put forward by Cahn (un-published). Fig. 7.6 shows the effect of a local reduction of lamellar spacing on solutions of the type A and B. For solution A, if such a reduction occurs there is an associated increase in ΔT with the result that a depression in the solid–liquid interface will fill in as it is overtaken by adjacent lamellae and so a net change in spacing will be established. (Note that lamellae tend to grow normal to the interface.) On the other hand a similar perturbation on solution B causes the area of decreased spacing to suffer a decrease in ΔT which results in a bump appearing on the interface which stabilises the spacing. Similar arguments for a local increase in spacing further shows that while B can be stable, A is inherently unstable. The choice of solution is therefore reduced to that in the neighbourhood of B.

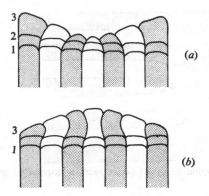

Fig. 7.6. Effect of local reduction of lamellar spacing on solutions of (a) type *A*, and (b) type *B*. 1 represents initial stable interface, 2 after applied perturbation, 3 the position of the interface having reacted to the perturbation (after Jackson & Hunt, 1966).

This modification of the choice of conditions does not affect the general form of the ΔT–λ–v dependence, and equations like those derived for the minimum (7.3), but with different constants are arrived at. Thus the new result is experimentally inseparable from the old unless all the relevant parameters are known sufficiently accurately to be able to differentiate between the constants of proportionality. An attempt to perform such a comparison for the lead–tin system was made by Jackson and Hunt but the results were inconclusive due to inadequate knowledge of the α–β interfacial free energy and the diffusion coefficients. However, more recent work by Jordan & Hunt (to be published) on the same system has shown that experimental results are consistent with a modified termination condition theory, such that if empirical values for the relevant constants are derived by inserting the experimental results for the λ–v^2 relationship into the theory, then the experimentally observed v–ΔT^2 relationship is accurately predicted with these parameters.

Returning to fig. 7.4, it is interesting to ask what would happen if the minimum in the curve for rods were below that for the lamellae rather than as drawn. In this case the construction used to give points *A* and *B* becomes impossible and we must deduce that a rod-like structure will be formed in preference to the lamellar one. This is the subject of the next section. In fact some recent discussions (Chadwick, 1968) have suggested that this inversion of the relative heights for the two minima is the normal situation rather than being a special circumstance so that the whole theory based on the condition for lamellar termination

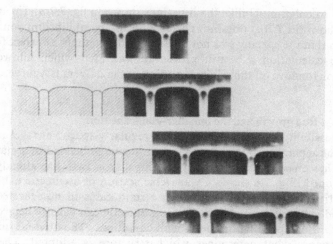

Fig. 7.7. Comparison of calculated and observed interface shapes in lamellar growth. Carbon tetrabromide–hexachloroethane eutectic, ×1300 (after Jackson & Hunt, 1966).

stability must be invalid. However, this assertion seems to be based on results (Moore & Elliott, 1968; Elliott & Moore, 1969) which contain errors and which are based on systems which do not correctly fall into the non-faceted/non-faceted category for which the theory was developed.

Before leaving the problem of lamellar growth it is interesting to note that the shape of the interface seems to be well described by the theories outlined above. Generally it is impossible to observe the shape of the solid–liquid interface during growth, particularly in the non-faceting (metal-like) materials, but using the organic analogue materials of Jackson and Hunt (see chapter 3) such observations become possible. Fig. 7.7 shows the shapes observed for a carbon tetrabromide–hexachloroethane eutectic with the calculated shapes matched on for comparison. In fact several empirical parameters including the lamellar spacing, the ratio of the lamellar thicknesses and the dihedral angle at the α–β – α–liquid – β–liquid boundary line were introduced into the calculation of the shape for this comparison. This obviously aids the closeness of the comparison considerably but the similarity between the micrographs and the calculated shapes is nevertheless quite striking. In fact these calculations of the shape were performed assuming that growth occurred under conditions given by the minimum of the curve for lamellae (see fig. 7.4) on the basis that this was simpler and should be sufficiently close to the conditions prevailing at point B for the purpose of this calculation. Unlike the situations shown in fig. 7.7(a), (b) and (c), that shown in 7.7(d) does not correspond to an equilibrium configuration in that the spacing was not the stable one, so that to correct for this in the calculation a best-fit procedure was used which allowed the solution to move off the minimum point on the ΔT versus λ curve.

7.3   Rod growth and the lamellar–rod transition

The possibility of a rod structure (i.e. regularly spaced parallel rods of one phase parallel to the growth direction and embedded in a matrix of the other phase) as an alternative to lamellar growth has already been mentioned. Fig. 7.8 shows a transverse section of a eutectic exhibiting this structure which is observed in certain cases in non-faceted/non-faceted eutectics, and indeed the same eutectic can usually be grown in either form depending on the growth conditions. The obvious question is therefore what determines which structure is adopted. The most natural criterion to apply is that of total free energy minimisation. Thus if a reduction in total interfacial free energy could be obtained by transforming a lamellar structure to a rod structure then this change should be anticipated. In order to examine this possibility it is necessary to know what the equilibrium spacings of lamellar and rod structures would be at a given growth velocity. (Evidently, on the basis of interfacial free energy alone, if one structure is much coarser than the other at the same growth rate, then this will be preferable due to its lower interfacial area.) Thus the problem reverts to considering again the curves of fig. 7.4 and the problem of selecting the relevant equilibrium points on the curves. Hunt & Chilton (1962) suggested that the lamella-to-rod transition should occur if the minimum on the rod curve is below the minimum on the lamellar curve; i.e. if the rods can grow at a lower undercooling. They found that this condition is satisfied when the volume fraction of the rod-like phase is less than $1/\pi(\sigma_L/\sigma_R)^2$ where σ_L is the average α–β interfacial free energy per unit area in the lamellar structure and σ_R is the same parameter for the rod-like structure. Thus for the case of isotropy of interfacial energy where $\sigma_L = \sigma_R$, the result predicts that eutectics having large volume fraction ratios will be rod-like whereas when the eutectic is fairly symmetrical, and the volume fraction ratio approaches unity, lamellae will normally be observed. This dependence on volume fraction is certainly found experimentally. The dependence on (σ_L/σ_R) also shows the main effect of marked anisotropy of the α–β interfacial free energy (which would probably be expected to exist in most systems). If we suppose that there is one strongly preferential plane for mutual orientation of the α- and β-phases

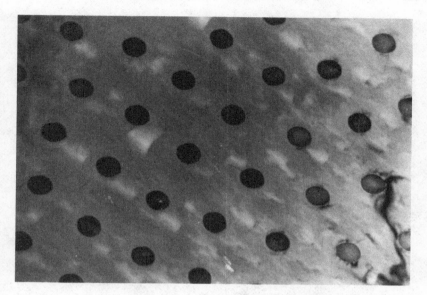

Fig. 7.8. Rod eutectic seen in transverse section: Al–Al$_3$Ni, × 20 000.
(Courtesy Dr M. H. Lewis & Dr H. W. Kerr.)

then if the growth conditions permit it, lamellae will be formed with this
interlamellar orientation. However, if growth is forced to occur in a
direction at a significant angle to this orientation, then the lamellar
interface has large areas in a high interfacial energy orientation and so
the rod structure may be preferred. Hunt and Chilton have shown that
this kind of effect can be demonstrated by growing eutectics round
corners and observing lamellar-to-rod transitions in the unfavourable
orientations.

An alternative approach to the understanding of the lamella-to-rod
transition was suggested by Chadwick (1963). He proposed that the
transition was a result of constitutional supercooling due to impurities
(i.e. the presence of a third species in addition to A and B). It is certainly
well known that constitutional supercooling can result in a cellular
structure (of cell size one or two orders of magnitude larger than the
interlamellar spacing). An example of this is seen in fig. 7.9 which shows
a transverse section of such a structure. A schematic drawing of the
growth of this type of structure is given in fig. 7.10 which shows the
solid–liquid interface shape and the lamellae within the cells. Indeed,
these cells were originally thought to represent separate grains and to
be an inherent feature of eutectic growth, and it was only the appreciation of cellular interface effects in single phase growth which led to an

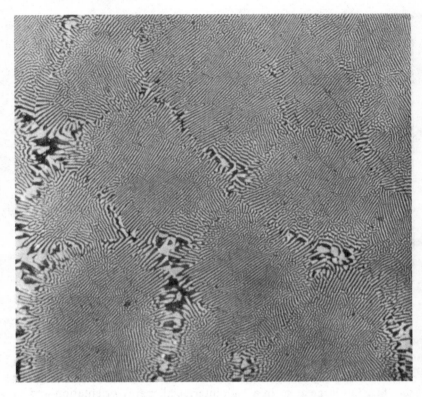

Fig. 7.9. Cellular eutectic structure in transverse section in a LiF–CaF₂ alloy, × 170. (Courtesy Dr M. W. A. Bright.)

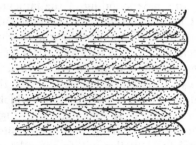

Fig. 7.10. Schematic drawing of growth occurring in the cellular lamellar eutectic mode showing the solid–liquid interface for *a* having a cellular form, and the orientation of lamellae in the cells which are growing normal in the interface.

understanding of these structures. Subsequently, the use of purer
materials allowed eutectics to be grown with an overall planar interface so that regular lamellar structures could be observed (as seen in
fig. 7.2). In many cases of cellular eutectic growth it is observed that
at the cell boundaries the lamellae break down into rods. It was this
observation that led Chadwick to propose the effect of impurity on lamellar breakdown since the cell boundaries will be the regions of greatest
impurity concentration. He proposed that, if the distribution coefficients
for the α- and β-phases relative to the added impurity are appreciably
different, the constitutional supercooling may be greater in one phase
than the other so that lamellae of this phase would break down. This
process is shown schematically in fig. 7.11. On the other hand, Hunt
(1966) has pointed out that this breakdown at the edges can be attributed
to the change in growth direction at the cell boundaries associated with
the curved interface. Indeed he points out that with different relative
orientations of the preferred lamellar orientation to overall growth
direction, cells may be almost all lamellar, all rod-like or a combination
of rods and lamellae, either symmetrically or assymmetrically.

Thus, as the Hunt and Chilton theory (subsequently confirmed by
the more complete analysis of Jackson and Hunt) seems to adequately
account for the observed effects there seems little virtue in adopting a
theory based on impurity effects which can only offer an alternative

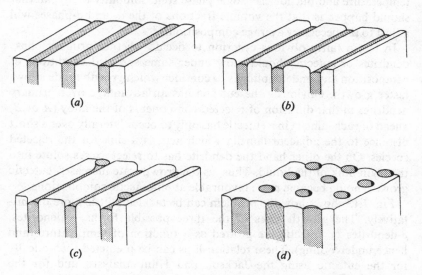

(a)

(b)

(c)

(d)

Fig. 7.11. Mechanism for lamellae to rod breakdown (after Chadwick
1963).

explanation in some of the cases observed. Moreover, each lamella is almost invariably growing into a region of constitutional supercooling due to the rejection of excess of one or other component of the alloy. It is therefore hard to understand why much smaller additions of a third species should radically affect individual lamellae (though an overall effect of constitutional supercooling leading to the cellular structure is much more clearly reasonable).

7.4   Eutectic range

In all the discussion so far it has been assumed that eutectic solidification occurs at the one singular point on the phase diagram, the eutectic point. If an alloy of some other composition is solidified then the phase diagram implies that the relevant primary phase should nucleate and in general will grow dendritically into the melt; if the liquid in between the dendrites becomes enriched so that it reaches the eutectic composition then eutectic growth can occur there. In uniaxial solidification this can take the form of lamellae or rods growing behind and between the primary phase dendrites (see for example, fig. 7.12). On the other hand, in the lamellar growth theory of Jackson and Hunt there is nothing in principle to stop eutectic growth (i.e. coupled growth of α- and β-phases) occurring anywhere below the eutectic horizontal (i.e. below the eutectic temperature and outside the single phase solid solubility limit); all that should happen is that the volume fractions of the α- and β-phases will adjust to the necessary average composition.

In fact a fairly obvious criterion to decide whether primary phase dendrites or eutectic should grow under some specified conditions of composition and undercooling is to consider which growth mode allows faster growth. Obviously the eutectic has an advantage over primary dendrites in that diffusion of rejected components of the alloy (*A* or *B*) ahead of each lamella in a eutectic has only to occur laterally over a short distance to the adjacent lamella which acts as a sink for the rejected species. On the other hand the dendrite has to reject excess solute into the main bulk of the liquid. Thus, as the growth rate increases, eutectic growth should become more favourable at off-eutectic compositions.

Fig. 7.13 shows how the problem can be tackled a little more quantitatively. The growth rates of the three possible forms, α-dendrites, β-dendrites or eutectic are plotted as a function of temperature (and hence undercooling). These relationships can be predicted theoretically for the eutectic using the Jackson and Hunt analysis, and for the dendrites using empirical relationships (in the absence of a theory for dendritic growth in alloys). Fig. 7.13(*a*) shows the situation for an alloy

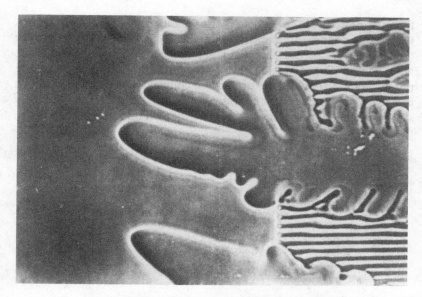

Fig. 7.12. Dendrites growing ahead of eutectic interface in a carbon tetrabromide-rich carbon tetrabromide–hexachloroethane alloy, × 125 (after Hunt and Jackson, 1967).

at the eutectic composition; all three mechanisms have the same equilibrium starting point, at the eutectic temperature T_E, but the eutectic can grow at the greatest rate at all temperatures. Fig. 7.13(b) then shows the situation for an alloy on the A-rich side of the eutectic point. Now at very low growth rates the α-primary phase dendrites should grow as expected because the equilibrium temperature (on the liquidus) is higher than for the eutectic (on the eutectic horizontal) or

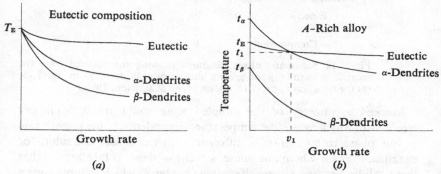

Fig. 7.13. Schematic growth rate versus temperature curves for eutectic and primary α- and β-phases at eutectic composition (a) and on the A-rich side of this composition (b) (after Hunt & Jackson, 1967).

138   *Eutectic growth*

for the β-primary phase (on some extrapolation of the β-phase liquidus).
On the other hand, for temperatures below T_1 or at growth rates greater
than v_1 the eutectic can once again grow faster despite the fact that the
alloy composition does not have the equilibrium eutectic value. The
phase diagram should therefore have a coupled growth zone as shown
in fig. 7.14(*a*). In this case the coupled zone is shown as approximately
symmetrical about the eutectic composition, but this need not always be
the case.

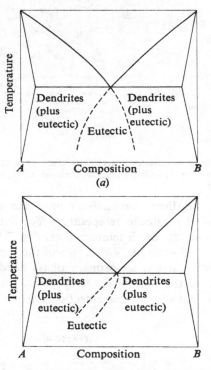

Fig. 7.14. Schematic phase diagrams showing the coupled zone for
eutectic growth: (*a*) for growth rate curves as depicted in fig. 7.13,
(*b*) for situation in fig. 7.15. (After Hunt & Jackson, 1967.)

Marked asymmetry of the coupled zone could result from any
mechanism which causes the temperature dependence of the growth rate
of one phase to be markedly different from the other. Consider, for
example, the case where one phase, say the α-phase, is faceting, so that
there will be a strong kinetic effect (see chapter 8) which will produce a
steep temperature dependence of the growth rate for the α-primary
phase and also, to a lesser extent, for the eutectic. In this case the growth

rate versus temperature curves may be as represented in fig. 7.15. (Note
that the α-primary phase is rather loosely referred to as dendritic; in fact
if the strong temperature dependence of its growth rate is because it is
faceting the morphology will not, of course, be dendritic.) Now at the
eutectic composition (fig. 7.15(a)) β-primary phase will grow at any
finite degree of undercooling, and of course this situation will be even
more enhanced on the B-rich side of the eutectic composition where the
β-equilibrium temperature will be higher than that for the eutectic or
the α-primary phase. However, on the A-rich side of the eutectic composition, α-primary phase must grow preferentially at low rates; at
higher rates the eutectic can grow most easily, whilst if the temperature
is reduced further, the β-dendrites win once again. Thus the asymmetric
coupled zone shown in fig. 7.14(b) is arrived at.

The foregoing arguments, which neglect temperature gradient effects,
are due to Hunt & Jackson (1967) following earlier discussion of the

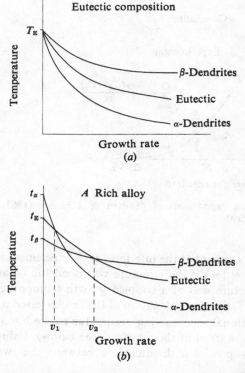

Fig. 7.15. As fig. 7.13 but with the α-phase having a strongly tempera-
ture-dependent growth rate (after Hunt & Jackson, 1967).

problem by Tammann and by Botschwar (1934). Experiments showing these effects have been performed by the latter authors and by Kofler (1950) amongst others. Recent experiments by Mollard & Flemings (1967) have shown that temperature gradients, as well as growth rates, can increase the eutectic range, possibly to any position in the phase diagram below the eutectic horizontal. Their experiments were performed on the tin–lead system giving the results shown in fig. 7.16. In this diagram (which is a plot of (G/v) versus the lead concentration,

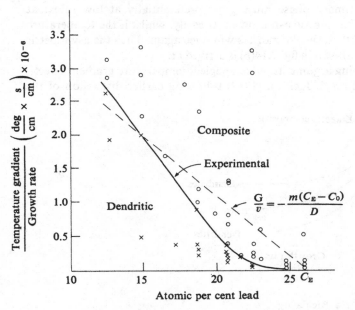

Fig. 7.16 Results of the experiments of Mollard & Flemings (1967). O composite, × dendritic.

where G is the temperature gradient at the interface) each experiment is represented either by a cross (×) if the structure was dendritic or an open circle (O) if the structure was of a coupled growth (composite) type. Mollard and Flemings offer an explanation of the results based on the argument that constitutional supercooling occurs due to the boundary layer which builds up in front of the interface (see below). Using this simple argument they predict a dividing line between the two growth modes given by

$$G/v = -m(C_E - C_0)/D \tag{7.4}$$

where m is the liquidus slope, C_E and C_0 are the eutectic and initial alloy compositions, and D is the liquid diffusion coefficient. Fig. 7.16 shows that this relation gives approximate agreement with the dividing line between the experimental observations.

The problem with applying arguments based on constitutional super-cooling to eutectics has already been mentioned in the previous section. For this reason the approach is rather dangerous and can be misleading; if it is desired to tackle the problem in this way, a proper stability analysis should be performed. Two attempts to do this have been made. Hurle & Jakeman (1968) have attempted to perform a detailed Mullins and Sekerka type of analysis (see chapters 5 and 6) on the lamellar eutectic structure described by the equations of Jackson and Hunt. They find that the structure will not be unstable to perturbations of wave-lengths much greater than the lamellar spacing, and so deduce that the instability leading to single-phase growth must be of shorter wavelength. Unfortunately, the steady-state solutions are not sufficiently well known to allow a proper analysis of the effects at these shorter wavelengths. Cline (1968) has attempted a much simpler calculation which predicts instabilities at long wavelengths (though no cellular structures were observed by Mollard and Flemings). However, his simplifications appear to miss out important factors which lead to the variation in conclusions (Hunt, Hurle, Jackson & Jakeman, 1970).

An alternative approach has been given by Jackson (1968) which is an extension of the foregoing discussion on effects at zero temperature gradient. Once again a comparison is made between the growth rates of the eutectic and primary phase dendrites, and the fastest growing form is assumed to win. Jackson and Hunt's analysis of the lamellar structure shows that, apart from the short range lateral diffusion which affects the solute concentration as a function of position on the interface, there is a boundary layer in which the average concentration varies normal to the interface. Perhaps rather surprisingly, it is found that the composition immediately ahead of the interface is always close to the equilibrium eutectic composition whatever the initial alloy composition. Fig. 7.17 therefore represents the variation of freezing temperature (relative to the primary phase most likely to grow out) as a function of distance from the interface, and the starting point of T_E will be approximately valid for all starting compositions. Also shown is the actual temperature for several possible temperature gradients (G_1, G_2, G_3). Clearly, for each temperature gradient there is a maximum degree of undercooling that a penetrating dendrite would experience (neglecting its own effect of disturbing the surroundings); for the case of zero gradient considered earlier this will occur far away from the interface (strictly at infinity).

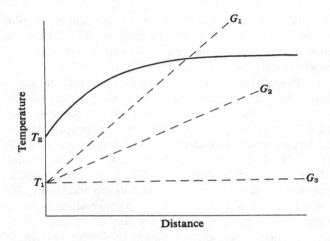

Fig. 7.17. Equilibrium temperature relative to one primary phase, and actual temperature for three different gradients ahead of growing lamellar eutectic (after Hunt, 1968).

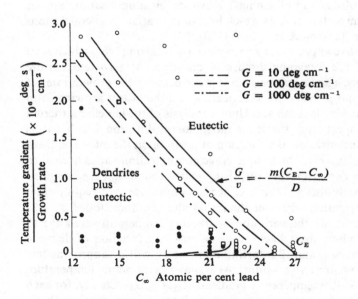

Fig. 7.18. Comparison of experimental results of Mollard & Flemings (1967) and theory of Jackson (1968). Observations of eutectic growth agreeing with the theory marked ○, agreements with dendrite observations are marked ● and disagreements are marked □.

For larger stabilising gradients this maximum is reduced. It is therefore possible to perform a modified analysis as before but taking as the dendrite undercooling this maximum value which will be a function of the temperature gradient. In this way Jackson has obtained a much improved correlation with the results of Mollard and Flemings (see fig. 7.18). Thus the eutectic range is a function of both (G/v) and of v.

7.5   Other structures

The lamellar and rod-like structures described so far are typical of non-faceted/non-faceted growth under controlled conditions and are by far the simplest eutectic structures observed. Indeed, not withstanding the difficulties apparent in the foregoing sections they are also the only structures which lend themselves to theoretical analysis.

The structures of eutectics involving a faceting phase are generally more complex but are qualitatively understood. Essentially, faceted/non-faceted eutectics can be classified roughly into three groups: rod-like and broken-lamellar structures which are fairly regular, irregular structures, and the so-called complex-regular structure. Irregular structures, as their name implies, do not have any obvious system or order in the eutectic morphology. Fig. 7.19 shows an example of such a structure in the aluminium–silicon system. Growth of this type can be observed using organic analogue materials and fig. 7.20 shows such a system during growth in which the faceting phase of borneol is seen growing as spikes which are covered (except at the tips) with the non-faceting phase of succinonitrile. The liquid left in the spaces between subsequently freezes at a lower undercooling into a finer structure. The regular rod-like structures and broken lamellae (see fig. 7.21) are similar to the non-faceted/non-faceted structures except that highly regular lamellae are not observed when one phase is faceting. These structures are usually observed when the faceting phase is the major volume fraction of the alloy. Finally the complex-regular structure seems to be observed in certain systems in which there is a slight excess of the faceting phase. The form of this structure can best be appreciated by studying fig. 7.22 which shows a transverse section of a complex regular structure in NaF–NaMgF$_3$ eutectic, and fig. 7.23 which shows the growth of such a structure in a cyclohexane–camphene eutectic. The structure seems to be one of lamellae growing in faceted cells. The three-fold symmetry observed in the transverse section of each 'cell' results from looking down on to overall facets forming the corner of a cube.

One feature which should be mentioned in relation to all these structures is that each phase is continuous. Sections of cast structures are

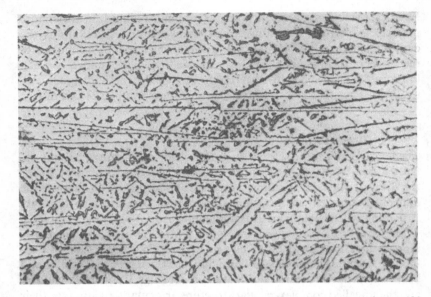

Fig. 7.19 Irregular structure in aluminium–silicon eutectic, ×430 (after Hunt & Jackson, 1966).

often misleading (e.g. fig. 7.21), in that they suggest that these less regular structures are comprised of isolated pieces of one phase embedded in a matrix of the other (higher volume fraction) phase. This would imply a need for continuous renucleation of the secondary phase. It was this misleading observation which led to one classification of eutectic structures as continuous or discontinuous. However, recent work has shown that many structures previously supposed to be discontinuous are almost certainly continuous. For this reason there is generally considerable doubt now that any truly discontinuous eutectic structures actually exist.

Hunt & Hurle (1968) have shown that these structures can qualitatively be understood in terms of the possible effects of facets of the single faceting phase. They argue that if facets can form on the faceting phase at the α–β–liquid groove then this can seriously impair the mobility of the grooves and prevent the structures from adapting themselves to fluctuations in the ambient conditions and to producing and maintaining regular lamellar or rod-like structures. Thus there will be a tendency for any regular structure formed in this way to be unstable and hence to break into some sort of irregular structure. On this basis it is only necessary to show why a breakdown does not occur in those cases

Fig. 7.20. Irregular structure growing in succinonitrile–borneol eutectic (from bottom to top), ×1300 (after Hunt & Jackson, 1966).

where the structure observed is regular. Firstly, the rod and broken-lamellar structures are almost invariably encountered when the faceting phase is the major volume fraction matrix phase. In this case it is probable that the faceting phase will indeed form facets and that the rods or broken-lamellae appear through depressions in these facets. As

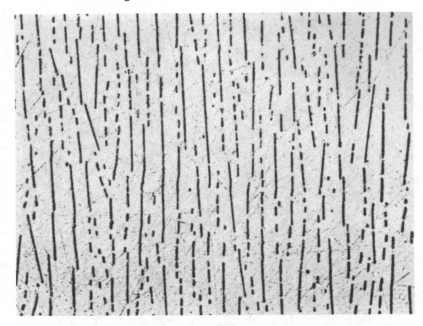

Fig. 7.21. Broken lamellar structure in bismuth–zinc eutectic. Transverse section × 120. (Courtesy Dr M. H. Lewis & Dr H. W. Kerr.)

Fig. 7.22. Complex regular structure in NaF–NaMgF$_3$. Transverse section × 100. (Courtesy Dr M. W. A. Bright.)

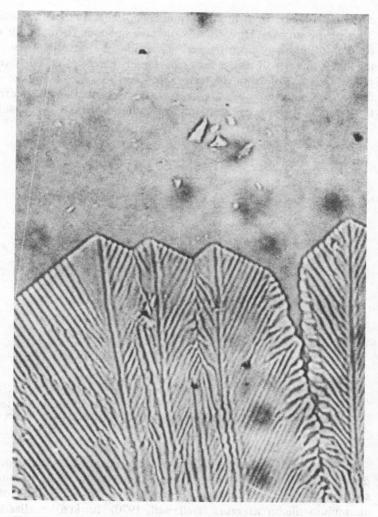

Fig. 7.23. Complex regular structure growing in cyclohexane–camphene eutectic (growth from bottom to top, × 1450) (after Hunt & Jackson, 1966).

faceting materials usually have only a small number of preferentially faceting planes spaced at wide angles apart it is not likely that a facet will form at the solid–liquid interface groove if the main planar interface is one of these facets. As fig. 7.24 shows, even if this interface does not correspond to one of the facet orientations the curvature of the faceting phase at the rod boundary must have one convex and one concave curvature and so, following the arguments of chapter 3, a facet cannot form in the groove. Thus these structures are able to maintain their regular morphology once formed. The same is true of the complex-regular forms. In this case the solid–liquid interfaces of the faceting phases over each face of the 'cells' are known to be part of the same facet plane. Thus no facets are likely to form in the boundaries around this large facet and hence the complex-regular structure will be stable.

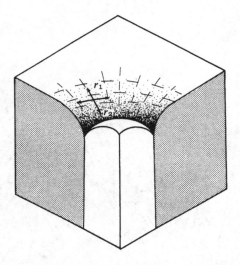

Fig. 7.24. Non-faceted rod or end of lamella emerging from faceting matrix showing form of curvature at groove (after Hunt & Hurle, 1968).

In many cases a eutectic alloy can adopt several possible structures depending on the growth conditions. For instance, small amounts of sodium impurity are known to produce a 'modification' of the aluminium–silicon structure (Hellawell, 1970). Broken lamellar structures often transform to rod-like structures with increasing growth rate (e.g. Elliott & Moore, 1969); in the past this transformation has been confused with a possible similar transition from regular lamellae to rods in non-faceted/non-faceted system.

Finally, the class of faceted/faceted eutectics seems to be much less interesting than the other systems. In many cases it appears that the two faceting phases grow at the same time but largely independent of one another and not in a coupled fashion. Fig. 7.25 shows an example of this type of growth occurring in an azobenzene–benzil eutectic.

Fig. 7.25. Azobenzene–benzil eutectic growing from bottom to top, ×270 (after Hunt & Jackson, 1966).

Further reading

There have been many reviews on eutectic growth in the past few years; in approximately chronological order these are Chadwick (1963), chapter six of Chalmer's book (1964), Kraft (1966), Kerr & Winegard (1966), Chadwick (1968), Hunt (1968) and more recently Hellawell (1970) and Hogan, Kraft & Lemkey (to be published). As the foregoing chapter may indicate, the subject is complicated and still changing quite rapidly. For this reason particularly, the earlier reviews in this list may well now be considered out of date. For instance Chadwick (1963) describes a classification of eutectic structures based on principles which he now accepts to be invalid (Chadwick, 1967). The last two reviews, by Hellawell and Hogan *et al.*, cover extensively certain aspects of eutectic growth largely neglected in this chapter; in particular, the structures of faceted/non-faceted eutectics and the crystallographic orientation relationships of the phases.

8 Mechanisms and kinetics of crystal growth

8.1 Introduction: two-dimensional nucleation

Undoubtedly one of the main reasons for wishing to understand the structure of solid surfaces and interfaces stems from the need to explain the processes of crystal growth. This can best be appreciated by first considering the three basic growth mechanisms which are thought to occur. Having established a model for the growth processes it is possible, by examining the kinetics of the process, to derive the relation between the velocity of advance of the interface and the thermodynamic driving force. For the case of growth from the melt the parameter which determines this driving force is the undercooling of the melt in the region of the interface (i.e. the difference in temperature, ΔT, between the actual interface and that of an interface of the same average curvature between phases of the same composition but which is static and in equilibrium). In principle this can be measured and related to the growth rates. It should, perhaps, be mentioned that many of the basic ideas contained in this chapter, especially in the first two sections, are derived from the studies and theories of growth from the vapour phase where their applicability is more soundly established.

Firstly then, we must start with a model of the interface. Let us assume for our present purpose that the interface is initially a singular (i.e. an atomically smooth) one. We have reason to believe that this may not be too unreal for some materials and it is certainly quite likely to be a reasonable description of a free solid surface (which is of interest in vapour growth). Suppose now the system is shifted from equilibrium by cooling so that atoms (or molecules) must condense from the liquid phase into 'solid atoms' on the interface.† The first atom that arrives

---

† We would generally suppose that solid–liquid equilibrium is a state in which atoms are arriving at, and leaving, the solid at the same average rate. This idea is specifically incorporated into theories of 'continuous growth' outlined in §8.3. However in theories involving an atomically smooth interface, different atom sites on the surface are supposed to be sufficiently different in their associated binding energies that at equilibrium atoms arriving at the interface from the liquid have such a low probability of sticking that they can be considered to never be incorporated in the solid. In this way the interface can be supposed to be truly static rather than static only in a time averaged fashion, and the idea of atoms only starting to arrive at the interface (for the purposes of solidification) when the equilibrium is disturbed, is a useful one.

may stick anywhere on the interface because all sites are equivalent and all involve the atom being weakly bound since most of its nearest neighbours are still in the liquid rather than the solid. The next atom, however, can choose either to take an isolated site similar to that occupied by the first, or it can take up a position adjacent to the first, in which case it (and the first atom) both have one more solid atom nearest neighbour than in the alternative configuration. For this reason this latter configuration is energetically favourable. Similarly, as subsequent atoms arrive they will also find it energetically favourable to adopt other sites adjacent to the previously deposited atoms. In this way the atoms arriving will build up monolayer 'islands' on the singular surface. Notice that these are likely to stay monolayers because it is just as difficult (i.e. energetically unfavourable) for an atom to lie alone on the island as on the initial surface (neglecting the existence of temperature gradients, of course). Thus growth will proceed by the spreading of these monolayers across the surface until they cover it, either by reaching the edge or running into another monolayer island. For growth to continue it is then necessary for one (or more) atoms to take up the initially highly unfavourable sites to start off another new layer. This process is essentially a two-dimensional nucleation and growth process.

An island on the surface contributes a small extra amount of interfacial free energy to the system due to its edge (the surface energy of its top being the same as that of the area of surface which it covers). This extra energy must be balanced against the reduction of free energy resulting from the transfer of atoms from the liquid to solid phase at a temperature below the melting point. The total extra free energy of an island containing n atoms can therefore be written as

$$\Delta G = -\alpha_1 n \Delta G_\mathrm{v} + \alpha_2 \sigma_\mathrm{a} \sqrt{n} \tag{8.1}$$

where α_1 and α_2 are shape factors, σ_a is the interfacial free energy per atom and ΔG_v is the volume free energy change per atom associated with the phase change. The problem can now be developed in the same manner as the three-dimensional case dealt with in §2.3. There will be a critical size for the island above which it can grow without increase in free energy. This occurs when it contains n* atoms given by

$$\sqrt{n^*} = \frac{\alpha_2}{\alpha_1} \frac{\sigma_\mathrm{a}}{2\Delta G_\mathrm{v}} \tag{8.2}$$

and taking ΔG_v to be $(L\Delta T/T_\mathrm{m})$ (with $L = \Delta H_\mathrm{f}$ as discussed earlier) gives

$$\sqrt{n^*} = \frac{\alpha_2}{\alpha_1} \frac{T_\mathrm{m}\sigma_\mathrm{a}}{2L\Delta T}. \tag{8.3}$$

This means that the critical value of the excess free energy associated with this island, or the barrier which must be overcome for growth to occur, is simply

$$\Delta G^* = \frac{\alpha_2^2}{4\alpha_1} \frac{\sigma_a^2 T_m}{L \Delta T}.$$ (8.4)

The rate of nucleation of monolayer islands can then be calculated. Following the arguments of chapter 2 and applying Boltzmann statistics, this will be proportional to $\exp(-\Delta G^*/kT)$.

It is now possible to calculate the growth rate. At very low nucleation rates, the process will occur as just described; that is, a single nucleation event will result in an island spreading across the surface to give a new layer. Then some time later, the next nucleation event will repeat the process. Under these conditions therefore, the growth rate depends only on the nucleation rate and not on the rate of arrival of atoms at the monolayer edge. Thus, the growth rate (i.e. velocity of advance of the interface) is

$$R = aI$$ (8.5)

where a is the thickness of each island monolayer and I the nucleation rate of islands and hence also of monolayers. As the rate of nucleation increases relative to the arrival rate of atoms at the ledges (at greater undercoolings), many nucleation events will contribute to the addition of each new layer and thus the rate of arrival of atoms and hence the rate of layer spreading must be considered explicitly in the growth rate. This introduces further dependence on ΔT, but for reasonably small ΔT these will be masked by the exponential term arising from the nucleation rate. Thus the dependence of growth rate on undercooling is essentially of the form

$$R \propto \exp\left(-\frac{\alpha_2^2}{4\alpha_1} \frac{\sigma_a^2 T_m}{L k T \Delta T}\right).$$ (8.6)

This means that for small undercoolings the growth rate will be negligibly small, but will increase rapidly as ΔT increases. At large undercoolings the nucleation rate will be so high that the rate of arrival will then be the limiting process and the growth is then best described by the 'continuous mechanism' which will be dealt with in §8.3.

In practice, while the two-dimensional nucleation type of behaviour may be attained in highly perfect crystals, growth is normally found to occur at observable rates even at the lowest undercoolings. That is, there would appear to be no nucleation barrier for growth to proceed.

This is even found for materials which facet in growth and which therefore seem most likely to fit the two-dimensional nucleation model. It was this observation in the growth of crystals from the vapour that led Frank in 1949 to point out the importance of imperfections, and in particular screw dislocations, in the growing crystal.

8.2   Screw dislocation growth mechanisms

The role of screw dislocations in the growth process is explained in detail in many basic texts on dislocations (see, for example, Cottrell (1953) or Read (1953)) as it was one of the greatest successes of dislocation theory. However, in view of its importance some description of the spiral growth mechanism would seem to be in place here. The ideas can best be appreciated by considering a schematic representation of a screw dislocation emerging at a singular surface. Fig. 8.1(*a*) shows such a situation for a crystal in which the atoms are represented by elastic cubes. It is seen that the surface is now a spiral ramp with a ledge where the ramp ends. Clearly this ledge forms a set of sites for easy deposition of new 'solid atoms'. The important feature, however, is that as atoms are added, the ledge does not move to the edge of the crystal surface to produce a complete layer, as with an island nucleus; instead it spirals round the dislocation line and continues to provide ledge sites no matter how many layers of atoms are added. The dislocation therefore provides a perpetual ledge and there is no longer any necessity to nucleate new layers to maintain growth. The rate of growth is determined only by the rate of arrival of atoms at the ledge and no free energy barrier has to be overcome.

Now if we consider all points on the ledge to be advancing at the same rate (i.e. due to a uniform rate of arrival of atoms) then due to the fact that the ledge is anchored at the point of emergence of the dislocation, it will develop into a spiral. Points father out on the ledge from the dislocation line have farther to travel to perform one revolution and hence complete a new monolayer (cf. the effect of equal runners on a laned racing track without the usual staggered start). This development is shown schematically in fig. 8.1. Thus the surface will become a spiral ramp during growth with the highest point being at the dislocation line.

Now it is evident that for a fixed rate of arrival of atoms at the surface, the fastest growth rate for the advance of the whole surface on average can be obtained by maximising the length of ledge per unit area; thus by tightening the spiral the ledge density and hence the growth rate is increased. There is, however, a limit to this tightening of the spiral as the curvature of the spiral at its centre must not exceed that of the

critical nucleus corresponding to the ambient undercooling. If it did
then the centre of the spiral would cease to grow. Thus the spiral has a
well-defined optimum pitch that it will attempt to adopt for any
particular undercooling.

We can now calculate the growth rate under this process. The spacing
between steps on the spiral surface will be of the order of $a\sqrt{n^*}$, where a

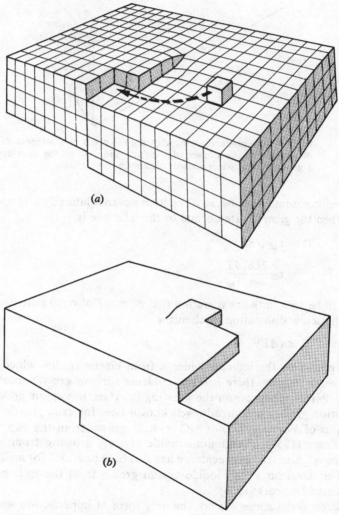

(a)

(b)

Fig. 8.1. (*continued overleaf*)

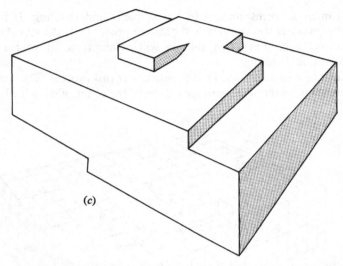

Fig. 8.1. Schematic drawing showing the point of emergence of a screw dislocation at the surface of a cubic solid, and of the development of a growth spiral when growth occurs on this surface.

is the interatomic spacing, so if the steps advance laterally with a velocity v_L, then the growth rate normal to the interface is

$$R = v_L/\sqrt{n^*}$$

$$= v_L \frac{2L\alpha_1 \Delta T}{\alpha_2 \sigma_a T_m}. \tag{8.7}$$

It will be seen in the next section that $v_L \propto \Delta T$ also, so that, according to the screw dislocation mechanism

$$R \propto (\Delta T)^2. \tag{8.8}$$

Apart from the indirect evidence from kinetic studies which will be discussed shortly, there is direct evidence for this growth mechanism. The spirals produced on the growing interface in vapour growth and solution growth are already well known (see, for example, the photographs of Verma (1951) of SiC crystals grown from the vapour, and of Forty (1951) of cadmium iodide crystals growing from aqueous solution). Similar direct evidence has also been provided for melt growth and a spiral on a lead iodide crystal grown from the melt has been reported by Sears (1953).

Screw dislocations are not the only form of imperfection which can assist crystal growth in this way, though they are probably the most common and usually the most important. For example, the re-entrants formed by twins intersecting a surface can also allow growth layers to form and spread without island nucleation. Re-entrants may be even more effective than screw dislocations when they are present because the limiting factor imposed by the critical radius on the pitch of a dislocation spiral does not apply to this case.

8.3   'Continuous' or 'normal' growth

The earliest ideas of crystal growth, due largely to Wilson (1900) and Frenkel (1932) were based on a rather different model. They assumed that all atoms arriving at the surface were able to stick to become solid. In other words, it was assumed that growth could equally well proceed from any point. This is the kind of behaviour which might be expected on a truly 'rough' interface, where, on average, all sites are equivalent. Thus the calculation of the growth rate simply requires the calculation of the rate of arrival of 'solid atoms' at the interface. This is clearly an oversimplification, particularly for the solid–liquid interface where we know that the condition of the interface equilibrium is really a continuous state of change. There will be continuous melting and freezing of material going on with atoms moving between the two phases near the interface; equilibrium simply means that the two rates are equal. Following the treatment of Jackson & Chalmers (1956) we can write the net rate of growth as

$$R = R_F - R_M,$$

the difference between the freezing and melting rates. These will each be processes having the usual exponential temperature dependence with activation energies for the two phases Q_L and Q_S, giving

$$R = R_F^0 \exp\left(-Q_L/kT\right) - R_M^0 \exp\left(-Q_S/kT\right). \tag{8.9}$$

Moreover, we have the condition that

$$Q_S - Q_L = L \tag{8.10}$$

and that at the melting point T_m, $R = 0$. Thus

$$\frac{R_F^0}{R_M^0} = \exp\left(-L/kT_m\right) \tag{8.11}$$

so that

$$R = R_F^0 \exp\left(-Q_L/kT\right)\left[1 - \exp\left(\frac{L}{kT_m} - \frac{L}{kT}\right)\right] \tag{8.12}$$

and, writing $\Delta T = (T_m - T)$, we have for reasonably small under-coolings that

$$(L\Delta T/kT_mT) \ll 1,$$

and therefore

$$R = R_F^0 \exp(-Q_L/kT)\frac{L\Delta T}{kT_mT}. \tag{8.13}$$

Thus, according to this mechanism of growth we have a rate of growth $R \propto \Delta T$ at small undercoolings. Since Q_L will be of the order of the activation energy for liquid diffusion, the term $\exp(-Q_L/kT)$ will be proportional to the liquid self-diffusion coefficient. This is inversely proportional to the viscosity, η, of the liquid so that for all undercoolings

$$R \propto \frac{1}{\eta}\left[1 - \exp\left(-\frac{L\Delta T}{kT_mT}\right)\right]. \tag{8.14}$$

8.4   A unifying theory

We have now considered three different possible mechanisms of growth. The natural reaction is to test them to see which predicts the right growth behaviour. If one is more satisfactory than the others then presumably this gives us a clearer picture of the nature of the interface and the growth mechanism. In fact it seems that for different materials all three mechanisms can operate. Returning to Jackson's (1958) theory as described in chapter 3 we can see that this might be expected. Low α-factor materials are believed to have rough interfaces and so the continuous growth mechanism would seem to be the relevant one. For high α-factor materials which are inclined to facet, a two-dimensional nucleation process is to be expected for the growth of highly perfect crystals with the screw dislocation mechanism becoming important for the lower driving forces (i.e. smaller undercoolings) when imperfections are present. Experiment largely bears this out although the results of experiments are often open to criticisms and are difficult to perform reproducibly. They do, however, show the trend anticipated.

If this view is accepted then there are essentially two different mechanisms of crystal growth. Firstly, there is the layer spreading theory in which the interface advances normal to itself only by the sideways motions of steps on the surface, the steps originating either from two-dimensional nuclei or from imperfections. The other, the continuous growth theory, allows the interface to advance everywhere concurrently.

The problem is to understand the borderline cases; some materials, having an α-factor of about 2 could behave in either fashion, depending on how diffuse the interface has to be before the continuous growth mechanism operates. Is it reasonable to suppose that there is a clearly defined borderline between the two cases? It seems more likely that there should be some gradual change from one mechanism to the other as the diffuseness of the interface changes. In other words, the two mechanisms are simply extreme forms of one general mechanism of growth.

In an attempt to unify these theories, Cahn (1960) proposed a phenomenological theory. His basic premises are, firstly, at sufficiently low growth rates the interface will always attempt to be in equilibrium, and secondly, the free energy of an interface is a periodic function of its mean position relative to the lattice periodicity of the solid phase. The first of these is fairly obvious; if the growth rate is sufficiently slow equilibrium must be maintained, and the only problem is deciding at what point this ceases to be true. The second is rather less obvious; certainly, if the interface advances by an integral number of monolayer spacings its free energy must be the same as before due to the translational symmetry of the lattice. It can then be argued that as each of these configurations has been defined as an equilibrium one, then states of higher free energy must have occurred in between. In fact, Temkin's analysis discussed in chapter 3 shows that this is so; there is always a minimum free energy configuration for the interface however diffuse, and the mean position of this interface occurs at some well-defined position relative to the atom layers. Thus the free energy of the interface as a function of position can be drawn schematically as in fig. 8.2. There is no reason, *a priori*, for assuming a sinusoidal form for the curve but this is a convenient way of representing the maximum and minimum free energy points on the displacement curves. (Note that even if this curve is not smooth, but has singularities in it, the ensuing argument is not affected.) On this basis a general theory of growth can be developed. If the applied driving force for growth is less than the free energy change between maxima in fig. 8.2 then the interface cannot pass through the corresponding intermediate positions. However, it can advance by a lateral mechanism in which the mean position of the interface over each terrace only differs from the next by an integral number of monolayer spacings (i.e. these are all minimum free energy configurations apart from the small ledge energy). On the other hand, if the driving force is sufficiently high the situation becomes more like fig. 8.3; there is no longer a potential barrier to continuous growth, which now becomes the most efficient way of advancing.

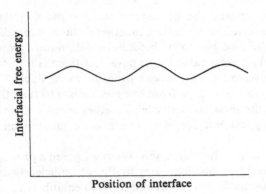

Fig. 8.2. Schematic graph of the free energy of an interface as a function of its mean position normal to the interface.

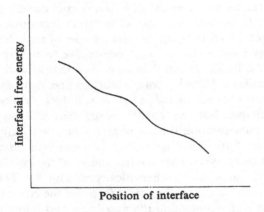

Fig. 8.3. Schematic graph of the free energy of an interface as a function of its mean position in the presence of undercooling in the melt.

Cahn has calculated this critical free energy barrier to be

$$\Delta G_V^* = \pi \sigma g / a \tag{8.15}$$

where σ is the interfacial free energy and g is a 'diffuseness parameter'. For a sharp interface $g \sim 1$ and for a very diffuse interface

$$g = \pi x^3 \exp(-\pi x) \tag{8.16}$$

where $x = \pi n/2$ and n is the number of atomic layers comprising the solid-to-liquid-transition at equilibrium. This means that the critical undercooling at which the growth mechanism changes from lateral to continuous is

$$\Delta T^* = \pi g T_m / aL. \tag{8.17}$$

In fact Cahn suggests that the lateral mechanism will hold for $\Delta T < gT_m/aL$, continuous growth for $\Delta T > \pi gT_m/aL$ with some transitional region at intermediate undercoolings.

Unfortunately, Cahn does not give a method of calculating the important g factor and so the theory can only be used to give an empirical value for g. Moreover, a conclusive experiment showing this transition occurring has yet to be performed. Certain results *suggest* that it may occur but they are far from indisputable. It is worth pointing out that in principle Jackson's theory allows for a transition to occur in borderline cases, since applying an undercooling to his calculations has a similar effect of tilting the excess free energy versus monolayer coverage curves (fig. 3.1) about the origin. Consequently, in materials having α-factors slightly greater than 2, the first minimum will move towards the 50 per cent coverage ('rough') position. The basic assumption of Cahn's is, however, fundamentally different as it suggests that *all* materials grow by a lateral mechanism at sufficiently low driving forces. A further criticism which has been made of Cahn's theory is that it is based on a second-order phase transformation and does not therefore apply to first-order transitions like solidification and melting (Jackson, Uhlmann & Hunt, 1967).

These problems are overcome in Temkin's paper of 1964, the basic model of which is discussed in chapter 3. Using his model of a solid–liquid interface, in which the thickness of diffuseness is a free parameter to be determined by minimising the free energy, he shows that at equilibrium the mean position of the interface is fixed relative to the lattice periodicity as already mentioned. He then goes on to consider the case of finite supercooling at the interface. The result is that he finds the solutions to the problem can fall into one of two types A or B depending on the values of two parameters β and γ (see fig. 8.4 which shows the curve separating these solutions). β is simply the driving force divided by kT so that

$$\beta \simeq \frac{L}{kT} \cdot \frac{\Delta T}{T_m} \tag{8.18}$$

and γ is related to the interfacial free energy such that

$$\gamma = \frac{4}{kT} [\epsilon_{12} - (\epsilon_{11} + \epsilon_{22})/2] \tag{8.19}$$

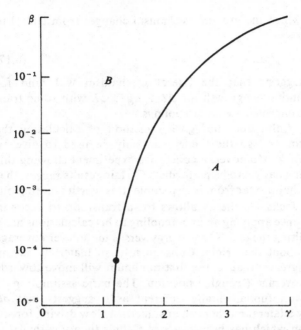

Fig. 8.4. Graph showing the two regions A and B as a function of parameters β and γ in which the two different solutions apply in Temkin's theory. After Temkin (1964).

where the ϵs are the bond energies (on a nearest neighbour bonding model) between adjacent atoms in the solid (phase 1) and liquid (phase 2). Note that for the simple cubic model used by Temkin, if the solid–liquid bond energy ϵ_{12} is considered equal to the liquid–liquid bond energy ϵ_{22} (as Jackson (1958) implicitly assumes) then $\gamma = \frac{2}{3}(L/kT)$. Thus γ is the same as (or similar to) Jackson's α-factor which would also be $\frac{2}{3}(L/kT)$ for the same (100) surface of a simple cubic material. Temkin finds that in region A there are the two solutions to the equilibrium problem as for zero undercooling; that is, one metastable and one stable, which is presumably the one of interest. Along the boundary between regions A and B these solutions merge, and in region B no equilibrium solution is present. Thus region A corresponds to a situation in which the well-defined equilibrium configuration occurs, so that a barrier to growth is present and the interface must advance by lateral spreading. In region B the absence of equilibrium configurations implies a 'barrierless' continuous growth mechanism.

We can now see why no transformation in the growth behaviour is normally observed. A typical metal has a value of $\gamma \lesssim 1$ so that the growth mechanism passes into the barrierless region B for $\beta \gtrsim 10^{-5}$ or less. Thus the transition occurs at undercoolings of the order of 10^{-2} deg or less. On the other hand, a typical faceting material, salol, with a γ of about 5, needs a relative undercooling $\Delta T/T_m > 0.1$ or more. Experiments to detect either of these effects are clearly not very feasible. A material with a γ of about 2 may be a rather better choice for this kind of experiment; such a material might be expected to undergo a transition from one type of growth to the other at an undercooling of about $10^{-2} T_m$. Such a system might therefore provide a transition at a reasonable growth rate and measurable undercooling.

8.5   Experiments on growth kinetics

It has already been mentioned that much of the experimental data on growth kinetics are subject to criticism. Both for this reason and because of the large volume of such work, which is of varying quality, no attempt will be made to review all such experiments. Rather the reader is referred to the existing literature for such information, and, in particular, to the reviews by Cahn, Hillig & Sears (1964) and Jackson, Uhlmann & Hunt (1967) which present similar material but from very different viewpoints. On the other hand some discussion of the techniques used together with examples of the results will serve to provide an appreciation of the difficulties encountered in such experiments and will serve to explain the rather unreliable state of current results.

Techniques.   Essentially, the techniques for experimentally obtaining information on the growth kinetics can be divided into direct and indirect techniques. The direct techniques simply involve direct measurement of both the interface temperature and the growth rate associated with it. Unfortunately, a direct measurement of the interface temperature is extremely difficult, partly because a measuring probe placed at the interface may well disturb the system, and partly because the temperature must generally be measured with great accuracy, because the quantity of interest is the *difference* between the interface temperature and equilibrium melting temperature ΔT and this is rarely greater than a few degrees and often only a small fraction of a degree. At higher undercoolings the growth rates are usually too high to be measurable or to allow a steady state of growth to be reached. A much more easily measured parameter is the temperature of the bath of large thermal mass in which the specimen is commonly placed in experiments on kinetics. This temperature can usually be measured with great accuracy. However, its use can lead to very misleading and confusing results, particularly under conditions of dendritic growth. Under these conditions it can be shown by considering the dendrite to be a paraboloid of revolution that the growth rate will be proportional to the square of the bath undercooling. However, this does not imply that a screw dislocation growth mechanism is operating; it is simply a result of the heat flow from the dendrite and shows that the growth has been limited by this. The only information which can be deduced from this result about the kinetics is therefore that the undercooling of the interface (due to kinetics) is very much less than that measured from the bath undercooling. However, despite all these limitations and because of the ease of measuring and accurately maintaining bath temperatures, this is often the approach used to obtain interface temperatures. If the experiment is performed with the specimen encased in thin walled capillary tubes in such a bath, then it is argued that the heat flow from the interface will maintain the interface at the bath temperature. In the limit this will certainly be true, but it is debatable whether sufficiently thin walled and sufficiently fine-bored capillaries are in fact used in these experiments. Also the approximation is only likely to be valid at low growth rates and hence very small undercoolings. For these reasons many results obtained in this way can be criticised for failing to take account of the finite temperature difference between the interface and the constant temperature bath, a difference which must be present to allow the dissipation of the latent heat of solidification. Obviously if this temperature difference is of the same order as the measured undercooling, the results of the experiments will be meaningless unless accurate corrections can be made for the thermal effects.

In order to make good use of the measurements of bath undercooling in kinetic studies the ideal material for such studies is obviously one with a very small kinetic coefficient; i.e. one which has very small growth rates for substantial undercooling. In these cases heat flow will no longer be the dominant consideration as there will now be ample time for heat dissipation, and indeed it may not even be necessary to use capillary tubes to contain the specimens. Materials which fall into this category are those having high viscosity melts near the melting point; this means that the diffusive motion of atoms in the melt is slow and so the net rate of arrival of 'solid atoms' at the interface is also slow. As an example of the small kinetic coefficients which may be encountered, a study of the crystallisation kinetics of sodium disilicate (Meiling & Uhlmann, 1967) finds growth rates of only about 10 microns per minute for an undercooling of 10 deg. This can be compared with growth rates of more like 5000 microns per minute for a 10 deg undercooling in a less viscous, typical faceting material like salol. For a metal such as tin a 10 deg undercooling (if obtainable) may well produce growth rates considerably in excess of 200 cm per minute. Evidently, therefore, the range of applicability of these direct techniques is limited. Moreover, high viscosity materials are usually the kind of substances which are difficult to obtain in high purity forms so that results obtained with them can always be subjected to criticisms relating to the impurity content. In vapour and solution growth, and probably in melt growth also, impurities are known to influence very seriously the observed growth kinetics; it is often argued, for example, that impurity atoms 'poison' ledge sites in lateral growth behaviour.

In addition to these direct approaches to the experimental problem, a rather novel indirect one known as the thermal wave technique was proposed by Kramer and Tiller in 1962. The principle is to equilibrate a solid–melt system in a fixed temperature gradient and then to impose on the hot (melt) end a small temperature variation which is sinusoidal in time. In this way a thermal wave is transmitted down the specimen, and causes the interface to advance and retard in a way which is a function of the solidification and melting kinetics. By studying the amplitude and phase of the thermal wave reflected from, and transmitted through the interface, it is possible, in principle, to determine the interface kinetics of the system. However, the analysis of the results is extremely complex and consequently the results are somewhat unreliable. This is demonstrated by the fact that two independent investigations on the kinetics of solidification of tin using the technique arrived at quite different conclusions; some of the problems involved in the interpretation of these experiments have been discussed by Sekerka (1967c). It has also been shown that the results can be misleading because of the possible effects of impurities (James & Sekerka, 1967) and because the peculiar conditions of the experiment (i.e. where cycles of melting and freezing are occurring) may produce non-typical kinetics, or a mixture of different kinetic conditions (O'Hara, Tarshis & Tiller, 1967). Thus, while the technique is both novel and elegant, it is by no means as useful as it first appears.

Some examples of results.   In order to appreciate the position reached in these studies of growth kinetics better, it is worth considering in detail the results of one or two experiments. For this purpose we will arbitrarily choose a metal and a high α-factor material. While we have seen that the α-factor may be only one of the factors determining kinetic mechanisms, these will provide useful extremes of behaviour. Because of the very high kinetic coefficients which appear to be typical for metallic growth, experiments on metals are scarce. For this reason direct techniques are quite unsuitable and the thermal wave technique offers almost the only method likely to give information for metals. Two studies using this technique have been made on tin, by Kramer & Tiller (1965) and by Rigney & Blakely (1966). The former workers concluded that there was a $(\varDelta T)^2$ dependence of growth rates in the range of undercoolings 6×10^{-3} to 6×10^{-2} deg, with associated growth rates between 2×10^{-5} and 2×10^{-3} cm s^{-1}. The latter workers concluded that there was a linear dependence in the range of undercoolings 2×10^{-3} to 0.5 deg with associated growth rates of 2×10^{-5} to 2×10^{-3} cm s^{-1}. The results could scarcely appear at greater variance. An interesting comparison can be made with the results of dendritic growth experiments performed on tin by Orrok (1958). He observed dendritic growth rates of 3 cm s^{-1} at a bath undercooling of 10 deg. As has already been pointed out, this bath undercooling will be very much greater than the interface undercooling due to the temperature gradients required for heat dissipation. Yet an extrapolation of Rigney and Blakely's results towards the growth rates measured by Orrok (admittedly, this is a rather excessive extrapolation) would imply that tin could not grow at 3 cm s^{-1} until the *interface undercooling* was of the order of 300 deg. Clearly the evidence is somewhat contradictory. Generally, this is true for most metallic and low α-factor materials. The kinetic coefficient is clearly large but the growth mechanism operating is not clear.

The faceting material which has probably received the most attention in the experimental evaluation of solidification kinetics is salol (phenyl salicitate). Some of the data for this material is represented in fig. 8.5 which is a plot of the 'reduced growth rate' versus undercooling. The reduced growth rate

$$\frac{R\eta}{[1 - \exp\{-(L\varDelta T/kT_mT)\}]}$$

is simply a way of correcting the growth rate for the effects of change in the rate of arrival of atoms at the interface as the temperature changes (see (8.14)). Thus, for continuous growth the reduced growth rate should be a constant for any material independent of undercooling, while for screw dislocation growth it should be linearly related to the undercooling. Thus the part of fig. 8.5 corresponding to undercoolings of less than about 46 deg corresponds quite closely to the behaviour predicted by the Cahn model of the transition from lateral growth to continuous growth in the undercooling range of about 6 to 16 deg, except that beyond this transitional region (undercoolings about 16 to 46 deg) the reduced growth rate is not truly constant but continues to rise slightly. Note that this value of the undercooling for the transition

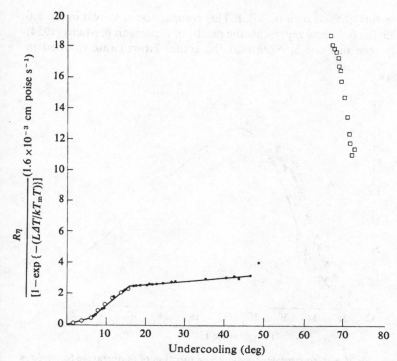

Fig. 8.5. Reduced growth rate versus undercooling for salol from the results of ○ Pollatschek (1929), ● Neumann and Micus (1954) and □ Malkin (1954) as presented by Jackson, Uhlmann & Hunt (1967).

is also rather less than that anticipated for salol by Temkin's analysis (see §8.4). However, the discrepancy of perhaps only a factor of 2 may not be significant in view of the uncertainty in making a suitable choice of Temkin's γ-factor. Danilov & Malkin (1964) produced results for salol (not represented on fig. 8.5) which showed negligible growth below about 1.6 deg undercooling followed by a sudden rise towards the values of Pollatschek (1929). This behaviour has been interpreted as evidence for two-dimensional nucleation and therefore it might be supposed that the crystals used in this investigation were of greater perfection than those in the other studies represented in fig. 8.5.

Jackson, Uhlmann & Hunt (1967) have questioned this simple and favourable interpretation of the results firstly by pointing out that a further break in the growth rate seems to occur at higher undercoolings (around 50 to 60 deg undercooling), which is not expected, and also by comparing the results of these other workers with those obtained from experiments of their own on salol. This comparison is shown in fig. 8.6 in which the full curve represents the results of Neumann & Micus (1954) already seen in fig. 8.5. Note that the actual growth rate (plotted in

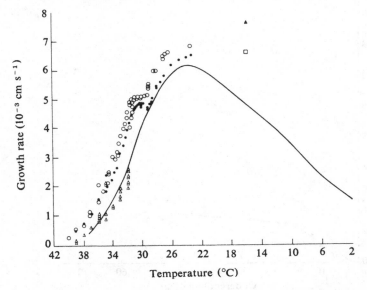

Fig. 8.6. Growth rates for salol as a function of temperature for various different conditions. After Jackson, Uhlmann & Hunt (1967).
△ degassed distilled rhombohedral face
● degassed as received
○ degassed distilled
□ first 8 mm of growth
▲ last 8 mm of growth.

fig. 8.6) passes through a maximum and then falls again at higher under-coolings due to the increase in viscosity of the melt. One of the main points made by Jackson *et al.* was that the purity of the material (including dissolved gases) could seriously affect the results of the investigation. This is shown to some extent by the scatter of experimental points, particularly at high undercoolings, for material treated in different ways (or at different stages of solidification which may also affect the impurity content because of segregation effects). Furthermore, all their growth rates were higher than those of Neumann and Micus and the turnover point in the growth rate at about 16 deg undercooling (around at temperature of 26°C) shown by the results of these authors is not evident in the new results. At low undercoolings (below about 10 deg) large single crystal rhombohedral plates grew; the scatter of

points at the same undercoolings for the open triangles in fig. 8.6 represent the variation in growth rates on different facets at one occasion
rather than a random scatter of several experiments. The kink in the
new results around 32°C (about 12 deg undercooling) corresponds to a
change in morphology as large facets broke up into smaller ones.
Finally, the authors commented that at low undercoolings the growth
rate of the crystals was sometimes observed to increase suddenly in
some directions by up to a factor of five for no apparent reason. Thus
their results emphasise the difficulties of carrying out reliable kinetics
experiments and, in particular, show the influence of impurities as well
as other unknown effects (possibly the generation of dislocations in the
crystals during growth) all of which can cause rapid fluctuations in the
growth rate. Their observations also show the importance of the morphology of the interface; changes in this can lead to quite significant
changes in the growth rate. These experiments therefore do not offer a
new solution to the problem, but rather question the simple interpretation of the earlier measurements.

While the lateral-to-continuous growth transition is therefore in
dispute, there does seem to be fairly decisive evidence for the existence
of a two-dimensional nucleation type of behaviour, obtained from a
rather different type of experiment. Various authors have reported a
transition from no growth to rapid growth which is characteristic of
two-dimensional nucleation behaviour. Care is needed in the interpretation of such evidence because in some cases at least, impurities have been
shown to have this same effect on growth rate. However, several pieces
of work report that, in systems where negligibly small growth rates are
observed at low undercoolings, a sudden rapid increase in the growth
rate can be induced by deforming the crystal, and hence introducing
dislocations. Some of the clearest indications of this behaviour are to be
seen in experiments on gallium which can be grown with relative ease in
a dislocation-free state, and which therefore provides a suitable material
for this kind of study (Pennington, 1966).

Evidence for a lateral-to-continuous growth transition. Some of the
evidence for, and against, the lateral-to-continuous growth transition
has been seen above in the salol system. Many other studies have been
made but they do not present any clearer evidence, at least, not in
favour of the transition. Some studies seem to produce results which are
fairly clearly contrary to the idea of a transition, but then the choice of
an unsuitable material might make the observation of the transition
almost impossible. For instance Meiling & Uhlmann (1967) have
measured growth rates and viscosities on the same samples of sodium

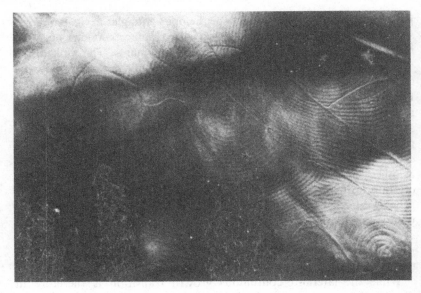

Fig. 8.7. Growth spirals on ice growing from water, × 75. Ketcham & Hobbs (1968).

disilicate (such a combined study is not available for salol) and find no evidence for the transition in the range of undercoolings 6 to 298 deg (the melting point of this material is 874°C).

However, it has been seen that kinetic data are generally not very reliable. It is therefore worth considering whether any other form of evidence for the lateral-to-continuous transition is available. One possibility is that the morphology of the interface may indicate the growth mechanism; in particular it might be supposed that lateral growth would be accompanied by a faceted morphology whereas continuous growth is typified by the absence of facets, and, at sufficiently large undercoolings, by the occurrence of dendritic growth. Thus the evidence that sodium disilicate grows with a faceted interface at an undercooling of 298 deg and salol grows with a faceted interface at 72 deg undercooling (Jackson *et al.*, 1967) may be taken to imply the absence of the transition in these materials up to these undercoolings. On the other hand, as pointed out earlier, the materials most likely to be suitable for observing a transition should be those having an α-factor (or Temkin α-factor) of about 2. This suggests that a study of the growth of ice might be useful. This is thought to have an α-factor of more than 2 and hence should give a faceted interface for growth normal to the basal plane (i.e. along the c-axis), whilst it should have a lower α-factor normal to this direction. Thus growth in the basal plane (normal to the c-axis) is observed at most undercoolings to be dendritic and no direct evidence of faceting has been found. The data on growth kinetics for ice are extensive but confusing, and therefore inconclusive. However, one interesting observation has been made by Ketcham & Hobbs (1968) who saw growth spirals (associated with lateral growth) on poly-crystalline ice specimens (see fig. 8.7). They suggest that the crystals in the polycrystalline sample will be randomly oriented and therefore that this is evidence of growth by a lateral mechanism at low undercoolings on surfaces parallel to the c-axis, which are known to grow dendritically at higher undercoolings. Providing their experimental system did not produce preferential orientations during nucleation so that all the crystallites were in fact growing on basal planes, this may be the most conclusive (albeit indirect) evidence of a transition so far.

8.6   Mechanisms and kinetics of melting

Few studies have been made of the melting of solids, but the experiments performed on kinetics, notably on high viscosity materials (for instance, sodium disilicate, phosphorus pentoxide, etc.) suggest that melting occurs by a continuous mechanism (i.e. a constant reduced growth rate). These materials facet during solidification and, as expected, display kinetic behaviour more typical of lateral growth. However, the apparent asymmetry of melting and solidification kinetics is probably not funda-mentally significant, but rather stems from the absence of facets in melting as a result of the particular experimental arrangements used, as discussed in chapter 3. Thus there is little evidence to help evaluate the kinetics and determine the mechanism of melting, and it can only be supposed that under suitable conditions to produce melting facets, some symmetry of melting and solidification kinetics might be expected.

Despite the lack of proper experiments designed to determine the interface kinetic some isolated observations on melting are relevant. In §2.5 it was pointed out that as the melts of most solids completely wet their parent solid at the melting point, superheating of solid surfaces is usually impossible. While the effects of impurities at free solid surfaces are always difficult to assess, several materials do not seem to fall into this category. In particular, it seems to be possible to superheat surfaces of crystals of p-toluidine (Sears, 1957) and gallium (Volmer & Schmidt, 1937) without producing melting. Indeed, Pennington (1966) reports that gallium does not appear to wet its solid at all, and that it can easily be shown to melt with a faceted interface without taking special pre-cautions (presumably a result of the different external surface effects

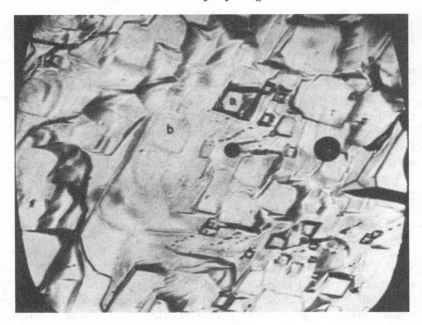

Fig. 8.8. Salol melting in a direction normal to the plane of view. × 70.

associated with the non-wetting behaviour). Moreover, it was seen in §2.6 that Mutaftschiev & Zell (1968) offer evidence for this non-wetting behaviour on the (0001) surface of cadmium by its melt, and it has been observed that zinc crystals 'skate' on their own melt, similarly suggesting a non-wetting behaviour. Unfortunately, while gallium facets easily in melting, attempts to evaluate the kinetics have been unsuccessful as the kinetic coefficient appears to be so large that no detectable superheating is present at the interface at manageable melting rates. However, Sears' (1957) experiments suggest that dislocations may play a significant role in the melting of these faceting materials. While he found he could superheat perfect faces of crystals of *p*-toluidine without melting, rapid melting occurred if this surface was then touched so as to induce local damage.

Finally, an observation of the author's on the melting of salol is shown in fig. 8.8. Salol crystals were melted under a microscope from the top to bottom face so that melting occurred along the microscopic axis and the plane of vision could be made the moving interface. Small diamond shaped features were seen to appear at many points on the interface and to enlarge rapidly as melting proceeded. These features are like that marked *b* on fig. 8.8. Pits like that marked *a* were much deeper and changed only slowly on melting. While no internal structure (such as that observed in growth spirals associated with screw dislocation growth) could be discerned in these features, they seemed reminiscent of dislocation etch pits. Certainly it seems likely that this behaviour was associated with the advance of the melting interface, and it is possible that dislocation pitting or even spiral mechanisms were the cause of the observations.

Further reading

The two reviews mentioned in the text (Cahn, Hillig & Sears (1964) and Jackson, Uhlmann & Hunt (1967)) provide a fairly extensive coverage of most of the theories and experimental work from rather different viewpoints. They also give reference to most of the relevant original papers which may be perused for further information. A review of the development of ideas relating to the effect of crystal imperfections, particularly in vapour phase growth was given by Frank in 1952.

References

Bethe, H.A. (1935) *Proc.Roy.Soc.* A150, 552.

Biloni, H. (1968) *Conference on the Solidification of Metals, Brighton, 1967*, Iron and Steel Institute, London, p. 74.

Biloni, H., Bolling, G.F. & Cole, G.S. (1966) *Trans.Met.Soc.AIME*, 236, 930.

Bolling, G.F. (1968) *Acta Met.* 16, 1147.

Botschwar, A.A. (1934) *Z.Anorg.Allg.Chem.* 220, 334.

Burton, W.K., Cabrera, N. & Frank, F.C. (1951) *Phil.Trans.Roy.Soc.* A243, 299.

Cahn, J.W. (1960) *Acta Met.* 8, 554.

Cahn, J.W. (1967) *Crystal Growth* (ed. H.S. Peiser), Pergamon, Oxford, p. 681.

Cahn, J.W. & Hilliard, J.E. (1958) *J.Chem.Phys.* 28, 258.

Cahn, J.W., Hillig, W.B. & Sears, G.W. (1964) *Acta Met.* 12, 1421.

Chadwick, G.A. (1963) *Progress Mat.Sci.* 12, 97.

Chadwick, G.A. (1968) *Conference on the Solidification of Metals, Brighton, 1967*, Iron and Steel Institute, London, p. 138.

Chalmers, B. (1959) *Physical Metallurgy*, Wiley, New York.

Chalmers, B. (1964) *The Principles of Solidification*, Wiley, New York.

Cline, H.E. (1968) *Trans.Met.Soc.AIME*, 242, 1613.

Coriell, S.R. & Hardy, S.R. (1969) *J.Appl.Phys.* 40, 1652.

Coriell, S.R. & Parker, R.L. (1967) *Crystal Growth* (ed. H.S. Peiser), Pergamon, Oxford, p. 703.

Cottrell, A.H. (1953) *Dislocations and Plastic Flow in Crystals*, Oxford University Press.

Cottrell, A.H. (1967) *An Introduction to Metallurgy*, Edward Arnold, London.

Danilov, V.I. & Malkin, V.I. (1964) *Zh.Fiz.Khim.* 28, 1837.

Davies, G.J. (1968) *Conference on the Solidification of Metals, Brighton, 1967*, Iron and Steel Institute, London, p. 66.

Davies, I.G. & Hellawell, A. (1969) *Phil.Mag.* 19, 1285.

Elcock, E.W. (1956) *Order–Disorder Phenomena*, Methuen, London.

Elliott, R. & Moore, A. (1969) *Scripta Met.* 3, 249.

Forty, A.J. (1951) *Phil.Mag.* 43, 72.

Forty, A.J. & Woodruff, D.P. (1969) *Techniques of Metals Research* (ed. R.F. Bunshah), Wiley, New York, vol. 2, p. 97.

Frank, F.C. (1949) *Disc.Faraday Soc.* 5, 48.

Frank, F.C. (1952) *Advances in Phys.* 1, 91.

Frank, F.C. (1958) *Growth and Perfection of Crystals* (ed. R.H. Doremus), Wiley, New York, p. 304.

Frank, F.C. (1963) *Metal Surfaces*, A.S.M., Metals Park, Ohio, p. 1.

Frenkel, J. (1932) *Physik Z.Sovjetunion*, 1, 498.

Geguzin, Ya.E. & Ovsharenkov, N.N. (1962) *Usp.Fiz.Nauk.* 76, 283. (English translation in *Sov.Phys.Uspekhi*, 5, 129.)

Gibbs, J.W. (1878) 'On the equilibrium of heterogeneous substances' in *Collected Works*, Longmans, Green and Co., New York (1928).

Glicksman, M.E. & Schaeffer, R.J. (1966) *Acta Met.* 14, 1126.

Glicksman, M.E. & Vold, C. (1969) *Acta Met.* 17, 1.

Gruber, E.E. & Mullins, W.W. (1967) *J.Phys.Chem.Solids*, 28, 875.

Hardy, S.C. & Coriell, S.R. (1968) *J.Crystal Growth*, 3,4, 569.

Hardy, S.C. & Coriell, S.R. (1969) *J.Crystal Growth*, 5, 329.

Hellawell, A. (1970) *Progress Mat.Sci.* **15**, 3.
Herring, C. (1951a) *Phys.Rev.* **82**, 87.
Herring, C. (1951b) *The Physics of Powder Metallurgy* (ed. W.E. Kingston), McGraw-Hill, New York, p. 143.
Hill, T.L. (1956) *Statistical Mechanics*, McGraw-Hill, New York.
Hilliard, J.E. and Cahn, J.W. (1958) *Acta Met.* **6**, 772.
Hogan, L.M., Kraft, R.W. & Lemkey, F.D., *Advances in Mat.Res.*, to be published.
Hulme, K.F. & Mullin, J.B. (1962) *Solid State Electron.* **5**, 211.
Hunt, J.D. (1966) *J.Inst.Met.* **94**, 125.
Hunt, J.D. (1968) *J.Crystal Growth*, **3, 4**, 82.
Hunt, J.D. & Chilton, J.P. (1962) *J.Inst.Met.* **91**, 338.
Hunt, J.D. & Hurle, D.T.J. (1968) *Trans.Met.Soc.AIME*, **242**, 1043.
Hunt, J.D., Hurle, D.T.J., Jackson, K.A. & Jakeman, E. (1970) *Met.Trans.* **1**, 318.
Hunt, J.D. & Jackson, K.A. (1966) *Trans.Met.Soc.AIME*, **236**, 843.
Hunt, J.D. & Jackson, K.A. (1967) *Trans.Met.Soc.AIME*, **239**, 864.
Hurle, D.T.J. & Jakeman, E. (1968) *J.Crystal Growth*, **3, 4**, 574.
Ivantsov, G.P. (1947) *Dokl.Akad.Nauk.SSSR*, **58**, 567.
Jackson, K.A. (1958) *Liquid Metals and Solidification*, ASM, Cleveland, Ohio, p. 174.
Jackson, K.A. (1965) *Ind.Eng.Chem.* **57**, 28.
Jackson, K.A. (1968) *Trans.Met.Soc.AIME*, **242**, 1275.
Jackson, K.A. (1971) *J.Crystal Growth*, **10**, 119.
Jackson, K.A. & Chalmers, B. (1956) *Can.J.Phys.* **34**, 473.
Jackson, K.A. & Hunt, J.D. (1965) *Acta Met.* **13**, 1212.
Jackson, K.A. & Hunt, J.D. (1966) *Trans.Met.Soc.AIME*, **236**, 1129.
Jackson, K.A., Uhlmann, D.R. & Hunt, J.D. (1967) *J.Crystal Growth*, **1**, 1.
James, D.W. & Sekerka, R.F. (1967) *J.Crystal Growth*, **1**, 67.
Jantsch, O. (1956) *Z.Krist.* **108**, 185.
Jones, D.R.H. & Chadwick, G.A. (1970) *Phil.Mag.* **22**, 291.
Kaykin, S.E. & Bené, N.P. (1939) *C.R.Acad.Sci.URSS*, **23**, 31.
Kerr, H.W. & Winegard, W.C. (1966) *J.Met.* p. 563.
Kerr, H.W. & Winegard, W.C. (1967) *Crystal Growth* (ed. H.S. Peiser), Pergamon, Oxford, p. 179.
Ketcham, W.M. & Hobbs, P.V. (1968) *Phil.Mag.* **18**, 659.
Kofler, A. (1950) *Z.Metallk.* **41**, 22.
Kotler, G.R. & Tarshis, L.A. (1968) *J.Crystal Growth*, **3, 4**, 603.
Kotler, G.R. & Tiller, W.A. (1967) *Crystal Growth* (ed. H.S. Peiser), Pergamon, Oxford, p. 721.
Kotler, G.R. & Tiller, W.A. (1968) *J.Crystal Growth*, **2**, 287.
Kotzé, I.A. & Kuhlmann-Wilsdorf, D. (1966) *Appl.Phys.Letters*, **9**, 96.
Kraft, R.W. (1966) *J.Met.* p. 192.
Kramer, J.J. & Tiller, W.A. (1962) *J.Chem.Phys.* **37**, 841.
Kramer, J.J. & Tiller, W.A. (1965) *J.Chem.Phys.* **42**, 257.
Kuhlmann-Wilsdorf, D. (1965) *Phys.Rev.* **140**, A1599.
Lindemann, F.A. (1910) *Phys.Z.* **11**, 609.
Lothe, J. & Pound, G.M. (1962) *J.Chem.Phys.* **36**, 2080.
Lothe, J. & Pound, G.M. (1968) *J.Chem.Phys.* **48**, 1849.
Malkin, V.I. (1954) *Zh.Fiz.Khim.* **28**, 1966.
Meiling, G.S. & Uhlmann, D.R. (1967) *Phys.Chem.Glasses*, **8**, 62.
Miller, W.A. & Chadwick, G.A. (1967) *Acta Met.* **15**, 607.
Miller, W.A. & Chadwick, G.A. (1968) *Conference on the Solidification of Metals, Brighton, 1967*, Iron and Steel Institute, London, p. 49.
Miller, W.A. & Chadwick, G.A. (1969) *Proc.Roy.Soc.* **A312**, 251.

Mollard, F.R. & Flemings, M.C. (1967) *Trans.Met.Soc.AIME*, **239**, 1526.
Moore, A. & Elliott, R. (1968) *Conference on the Solidification of Metals, Brighton, 1967*, Iron and Steel Institute, London, p. 167.
Morris, L.R. & Winegard, W.C. (1967) *J.Crystal Growth*, **1**, 245.
Muller, E.W. & Tsong, T.T. (1969) *Field Ion Microscopy*, American Elsevier, New York.
Mullins, W.W. (1959) *Acta Met*. **7**, 746.
Mullins, W.W. & Sekerka, R.F. (1963) *J.Appl.Phys*. **34**, 323.
Mullins, W.W. & Sekerka, R.F. (1964) *J.Appl.Phys*. **35**, 444.
Mutaftschiev, B. & Zell, J. (1968) *Surface Sci*. **12**, 317.
Nakaya, U. (1956) *Research Paper 13*, Snow, Ice and Permafrost Research Establishment, Corps. of Engineers, US Army, Wilmette, Illinois.
Nason, D. & Tiller, W.A. (1971) *J.Crystal Growth*, **10**, 117.
Neumann, K. & Micus, G. (1954) *Z.Physik.Chem*. **2**, 25.
Nicholas, J.F. (1968) *Austral.J.Phys*. **21**, 21.
O'Hara, S., Tarshis, L.A. & Tiller, W.A. (1967) *J.Chem.Phys*. **46**, 2800.
Oldfield, W. (1968) *Conference on the Solidification of Metals, Brighton, 1967*, Iron and Steel Institute, London, p. 70.
Onsager, L. (1944). *Phys.Rev*. **65**, 117.
Onsager, L. & Kaufmann, B. (1946) *Report on International Conference on Fundamental Particles and Low Temperatures, Cambridge, 1946*, Physical Society, vol. 2.
Orrok, G.T. (1958) Ph.D. thesis, Harvard University.
Papapetrou, A. (1935) *Z.Krist*. **92**, 89.
Pennington, P.R. (1966) Ph.D. thesis, University of California, Berkeley.
Pfann, W.G. & Hagelbarger, D.W. (1956) *J.Appl.Phys*. **27**, 12.
Pollatschek, H. (1929) *Z.Physik.Chem*. **142**, 289.
Read, W.T. Jr (1953) *Dislocations in Crystals*, McGraw-Hill, New York.
Rhines, F.N. (1956) *Phase Diagrams in Metallurgy; their Development and Application*, McGraw-Hill, New York.
Rigney, D.A. & Blakely, J.N. (1966) *Acta Met*. **14**, 1375.
Rutter, J.W. & Chalmers, B. (1953) *Can.J.Phys*. **31**, 15.
Sears, G.W. (1955) *J.Chem.Phys*. **23**, 1630.
Sears, G.W. (1957) *J.Chem.Phys.Solids*, **2**, 37.
Seidensticker, R.G. (1967) *Crystal Growth* (ed. H.S. Peiser), Pergamon, Oxford, p. 733.
Sekerka, R.F. (1965) *J.Appl.Phys*. **36**, 264.
Sekerka, R.F. (1967*a*) *Crystal Growth* (ed. H.S. Peiser), Pergamon, Oxford, p. 691.
Sekerka, R.F. (1967*b*) *J.Phys.Chem.Solids*, **28**, 983.
Sekerka, R.F. (1967*c*) *J.Chem.Phys*. **46**, 2341.
Sekerka, R.F. (1968) *J.Crystal Growth*, **3, 4**, 71.
Skapski, A.S. (1956) *Acta Met*. **4**, 576.
Spittle, J.A., Hunt, M.D. & Smith, R.W. (1964) *J.Inst.Met*. **93**, 234.
Stowell, M.J. (1970) *Phil.Mag*. **22**, 1.
Tammann, G.T. (1925) *States of Aggregation*, Van Nostrand, New York, chapter 9.
Taylor, M.R., Fidler, R.S. & Smith, R.W. (1968) *J.Crystal Growth*, **3, 4**, 666.
Temkin, D.E. (1960) *Dokl.Akad.Nauk.SSSR*, **132**, 1307.
Temkin, D.E. (1964) *Crystallisation Processes* (ed. N.N. Sirota, F.K. Gorskii and V.M. Varikash). English translation published by Consultants Bureau, New York, 1966.
Temkin, D.E. (1966) *Growth and Imperfections of Metallic Crystals* (ed. D.S. Ovsienko). English translation published by Consultants Bureau, New York, 1968, p. 11.

Tiller, W.A. (1958) *Liquid Metals and Solidification*, ASM, Cleveland, Ohio, p. 276.
Tiller, W.A., Jackson, K.A., Rutter, J.W. & Chalmers, B. (1953) *Acta Met.* 7, 428.
Turnbull, D. (1949) *J.Appl.Phys.* 20, 817.
Turnbull, D. (1950a) *J.Chem.Phys.* 18, 768.
Turnbull, D. (1950b) *J.Appl.Phys.* 21, 1022.
Turnbull, D. & Cech, R.E. (1950) *J.Appl.Phys.* 21, 804.
Turnbull, D. & Fisher, J.C. (1949) *J.Chem.Phys.* 17, 71.
Turnbull, D. & Hollomon, J.H. (1951) *Physics of Powder Metallurgy* (ed. W.E. Kingston), McGraw-Hill, New York, p. 109.
Tyndall, J. (1858) *Proc.Roy.Soc.* 9, 76.
Ubbelohde, A.R. (1965) *Melting and Crystal Structures*, Oxford University Press.
Udin, H. (1951). *Trans.AIME*, 197, 63.
Uhlmann, D.R. & Chalmers, B. (1965) *Ind.Eng.Chem.* 57, 18.
Uhlmann, D.R., Chalmers B. & Jackson, K.A. (1964) *J.Appl.Phys.* 35, 2986.
Verhoeven, J.D. & Gibson, E.D. (1971) *J.Crystal Growth*, 17, 29 and 39.
Verman, A.R. (1951) *Phil. Mag.* 42, 1005.
Volmer, M. & Schmidt, O. (1937) *A.Phys.Chem.* 85, 467.
Vonnegut, B. (1948) *J.Colloid Sci.* 3, 563.
Weatherley, G.C. (1968) *Met.Sci.J.* 2, 25.
Wilson, H.A. (1900) *Phil.Mag.* 50, 238.
Woodruff, D.P. (1968a) *Phil.Mag.* 17, 283.
Woodruff, D.P. (1968b) *Phil.Mag.* 18, 123.
Woodruff, D.P. & Forty, A.J. (1967) *Phil.Mag.* 15, 985.
Wulff, G. (1901) *Z.Krist*, 34, 449.
Zadumkin, S.N. (1962) *Fiz.Metal.Metalloved*, 13, 24. (English translation in *Phys.Metals Metallography* p. 22.)
Zener, C. (1946) *Trans.AIME*, 167, 550.

Index

Numbers in **bold face** refer to the first page number of a section or chapter relating to the subject. Ranges of numbers (e.g. 15–20) are used when reference is made to each of this inclusive range of page numbers.

Index